Geographies of Plague Pandemics

Geographies of Plague Pandemics synthesizes our current understanding of the spatial and temporal dynamics of plague, *Yersinia pestis*. The environmental, political, economic, and social impacts of the plague from Ancient Greece to the modern day are examined. Chapters explore the identity of plague DNA, its human mortality, and the source of ancient and modern plagues. This book also discusses the role plague has played in shifting power from Mediterranean Europe to north-western Europe during the 500 years that plague has raged across the continent. The book demonstrates how recent colonial structures influenced the spread and mortality of plague while changing colonial histories. In addition, this book provides critical insight into how plague has shaped modern medicine, public health, and disease monitoring, and what role, if any, it might play as a terror weapon.

The scope and breadth of *Geographies of Plague Pandemics* offers geographers, historians, biologists, and public health educators the opportunity to explore the deep connections among disease and human existence.

Mark Welford is a nature-society geographer at Georgia Southern University. His research interests include: environmental change in, and conservation of, tropical montane environments; hurricanes and climate change; and the spatial dynamics of historical pandemics. He has taught at Georgia Southern since 1993. He has also directed Study Abroad trips to Ecuador, India, the Czech Republic, Poland and Italy.

First published 2018 by Routledge

2 Park Square, Milton Park, Abingdon, Oxfordshire OX14 4RN
52 Vanderbilt Avenue, New York, NY 10017

Routledge is an imprint of the Taylor & Francis Group, an informa business

First issued in paperback 2020

British Library Cataloguing-in-Publication Data
A catalogue record for this book is available from the British Library

Library of Congress Cataloging-in-Publication Data
A catalog record for this book has been requested

ISBN: 978-1-138-23427-7 (hbk)
ISBN: 978-0-367-59241-7 (pbk)

Typeset in Times New Roman
by Sunrise Setting Ltd, Brixham, UK

Contents

Figures

Acknowledgments

I would like to thank Brian Bossak for helping me develop an academic interest in plague, Ian Blackburn for his cartography, Caren Town for copyediting the book, and Terry for support and understanding.

1 Plague, its emergence and persistence through recent human history

The bacterium, *Yersinia pestis*, known as plague, has killed tens if not hundreds of millions of people over the last 2,000–2,500 years and yet has been responsible for some of the most critical innovations in public health – those of quarantine and disease surveillance. Plague also helped end feudalism by inflating labor wages, and possibly seeded the transfer of economic power from Mediterranean Europe to north-west Europe during the medieval period by decimating the trading super-powers of Venice, Genova and Florence. Plague victims were also catapulted into Kaffa in 1346 by the Mongols under Janibeg in an attempt to break the siege and therefore represent the first use of biological terror weapons.

Unlike HIV-1 or SARS or Ebola that emerged in the 20th century, *Yersinia pestis*, commonly known as plague, diversified from *Yersinia pseudotuberculosis* – a common, environmental stress-tolerant, less pathogenic, enteric bacterium – between 2,600 and 28,000 years ago (Achtman *et al.* 1999, 2004; Cui *et al.* 2013; Wagner *et al.* 2014). In contrast to *Y. pseudotuberculosis*, the modern bacterium *Y. pestis* exhibits three forms that vary in their pathogenicity and transmissibility and their ability to become highly lethal, pandemic contagions: bubonic (BP), pneumonic (PP) and septicemic plague (SP) (Perry and Fetherston 1997). It appears that the emergence of the common ancestor of all *Yersinia pestis* strains occurred roughly 5,783 years ago (Rasmussen *et al.* 2015). Here 'pathogenicity' simply refers to an organism's ability to cause disease and harm while 'transmissibility' refers to the organism's ability to move from person to person or host to host across space and time.

Since diversifying from its common ancestor, *Yersinia pestis* has killed tens of millions through the First Pandemic (from 541–715 CE), the Second Pandemic (from the 1330s to 1879 CE) and the Third Plague Pandemic (from 1894 CE to the present). CE refers to the *common era*, that is, since AD 1, while BCE is *before the common era*. Historians suggest a third of the Roman Empire succumbed to the Justinianic Plague within the 6th century, and from 541–715 CE Europe lost as much as 50–60% of its population, between 30–50 million people. From 1347–1353 some 30–40% of the European population died, as many as 28 million people (Perry and Fetherston 1997). The beginning of the Second Pandemic is couched in Eurocentric terms; it is the beginning of the medieval Black Death, otherwise known as the primary wave of the medieval Black Death from 1347–1353. However, the plague was killing people in Central Asia and along the Silk Road/ Network at least ten years before its transmission into Europe (Schmid *et al.* 2015). In the latter half of the 14th century, China lost 25–30% of its population due to plague (Sussman 2011). Europeans, or more specifically Genoese traders, were also first exposed in Kaffa, part of the Crimea, in 1346 (Wheelis 2002), although plague did not arrive in Europe until 1347.

It is estimated that in the subsequent waves of medieval Black Death, between 1361 and 1879, fully a third of Europeans died. Just in Messina alone, in 1743, some 48,000 people died of plague (Cohn 2008). During the third pandemic, between 1898 and 1918, it is estimated that 12.5 million people in India died (Perry and Fetherston 1997). It is quite conceivable that plague has killed between 70–200 million people, making *Yersinia pestis* without a doubt the deadliest bacterium humankind has ever encountered.

The extent and lethality of these three pandemics appear to be linked to human relations to the natural world and connectivity between fellow humans through trade and transportation. Rather surprisingly, genomic analysis of modern and medieval plague has found no difference in the virulence-associated genes that control plague pathogenicity (Bos *et al.* 2011). Bubonic plague pathogenicity is a product of several processes: a plasminogen activator (Pla) that stops blood-clotting, thus aiding dispersal within victims and hosts; the Yops protein on the outer bacterium shell and the pH 6 antigen that trigger cytotoxic (hypersensitive allergic) reactions and immune suppression; an F1 capsular antigen that suppresses white blood cell phagocytosis (which stops white blood cells ingesting the plague bacterium); and an endotoxin that triggers septic shock and inflammatory response syndrome (Gage and Kosoy 2005). So, although plague's virulence has changed little over time, human relationships with nature and our connectivity to other humans has changed over time, and so geographers discuss diseases based largely on their spatio-temporal characteristics. Those diseases that occur among a defined population or within a defined spatial boundary or region are described as 'endemic diseases' (Bossak and Welford 2010). In contrast, 'pandemic diseases' occur over the largest spatial domains, frequently causing worldwide infection. Moreover, endemic diseases typically persist in a region; epidemics occur across multiple regions, while pandemics are generally highly transmissible and lethal across broad spatial domains (Bossak and Welford 2010). However, a pandemic disease can spread widely across the planet, yet not cause significant mortality – e.g., the swine flu of 2009 (Smith *et al.* 2009). In other words, transmissivity or infectivity is not necessarily indicative of severity (Bossak and Welford 2010).

What is plague?

Today, the most common, less lethal form of plague is bubonic plague; it is a necessary prerequisite for septicemic plague but not necessarily for pneumonic plague (Bos *et al.* 2016). Between 1947 and 1996, 84% of plague cases were bubonic, while 13% were septicemic plague, and just 3% were pneumonic plague (Bellazzini and Myint 2006). Nevertheless, bubonic plague is not easily trans-mitted between humans; it struggles to cross from the bloodstream into the lung alveoli, and it is not easily transmitted through aerosol or blood droplet forms (Kool and Weinstein 2005; Bossak and Welford 2009). In fact, bubonic plague requires flea vectors to transmit the bacterium from infected black rats or ger-bils or marmots to humans (Figure 1.1). However, for fleas to leave their hosts and seek a blood-meal elsewhere, the hosts must die. If these fleas seek blood-meals from humans, humans typically represent dead ends for the bacterium, as humans are very susceptible to plague and exhibit high mortality but rarely trans-mit bubonic plague. Only in Madagascar and Tanzania have human fleas sup-ported domestic human-to-human transmission of bubonic plague (Ratovonjato *et al.* 2014).

Septicemic plague is the blood infection form of bubonic *Yersinia pestis*. It has an identical modern case fatality rate (50–90%), and manifests in lymph nodes as buboes. In contrast, pneumonic plague begins as an infection in the lungs and spreads through the coughing of aerosol droplets (Bellazzini and Myint 2006). However, modern untreated pneumonic plague case fatalities vary between 95–100% (Inglesby *et al.* 2000). Today, all three forms of *Yersinia pestis* are rarely encountered. This is likely due to its low basic reproduction number (R_0), modern disease surveillance systems, widespread availability of cheap effec-tive plague antibiotics, and effective quarantine practices. R_0 is defined as the

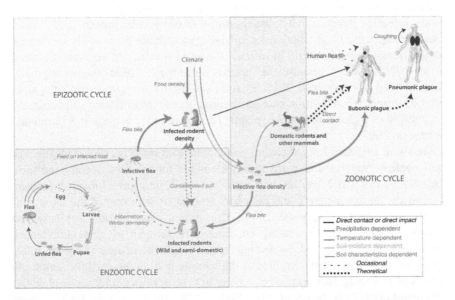

Figure 1.1 Plague cycle with small mammals as hosts and fleas as vectors. (Reprinted with permission from T.B. Ari, S. Neerinckx, K.L. Gage, K. Kreppel, A. Laudisoit, H. Leirs and N.C. Stenseth, 2011. Plague and climate: scales matter. *PLoS pathogens*, 7(9), p.e1002160.)

number of secondary infections that result from one primary infected individual. For instance, pertussis (whooping cough), a highly infectious disease, has a R_0 = 12–17, meaning that each primary host (each human) infects 12 to 17 other humans. In comparison, *Yersinia pestis* appears to have a R_0 between 1–3.5 (Begier *et al.* 2006; Kool and Weinstein 2005; Nishiura *et al.* 2006; Seal 1960). Outbreaks of pneumonic plague yield basic reproduction numbers that vary from a R_0 of 0.96 to 2.3 with an average of 1.3 (Gani and Leach 2004). In a very understated way, Gani and Leach (2004) note that high values of R_0 (up to and including 2.3) could generate large outbreaks that would be difficult to control. To put this R_0 of 2.3 in context, the Spanish flu of 1918–1920 (with an average R_0 of 1.8) killed somewhere in the order of 50 million people (Johnson and Mueller 2002). This at a time of great global social and political upheaval, when massive numbers of people and military personnel were moving across the globe following World War I. Yet this pandemic also occurred when quarantine practices were effective and when modern medicine, though in its infancy, understood what germs were, how they were transmitted, and how to combat pandemic outbreaks. However, R_0 values calculated for the deadly 1918–1920 Spanish influenza pandemic do vary between a R_0 of 1.8 (Biggerstaff *et al.* 2014) and 2.2–3.5 (Chowell *et al.* 2007).

On the other hand, close contact between humans infected with modern strains of pneumonic plague is necessary for transmission (Gani and Leach 2004). During the pneumonic plague epidemic that erupted in 1910–1911 among marmot hunters in Manchuria, the disease moved 965 km in three weeks in overcrowded and poorly ventilated railway cars (Chernin 1989), but once patients reached hospitals, infection transmissivity abated rapidly (Kool and Weinstein 2005). More recent pneumonic plague outbreaks in Madagascar have been easily contained, as few *Yersinia pestis* bacilli have been found in respiratory aerosols (Simonet *et al.* 1996). This is because the modern strain of bubonic plague originally derived from rats (Bos *et al.* 2016; Spyrou *et al.* 2016) struggles to cross the alveoli. So, although human-to-human pneumonic plague transmission can occur, the probability of a global pneumonic plague pandemic today is quite low (Kool and Weinstein 2005).

Genomic evidence from skeletons associated with mass burials of plague victims confirm that *Yersinia pestis* was responsible for the first, second and third pandemics. Prior to 2004, *Yersinia pestis* was split into three biovars (*Antiqua*, *Medievalis* and *Orientalis*), and each was associated with one of the three pandemics (Devignat 1951; Varlik 2008). This biovar classification was based on plague's ability, under laboratory conditions, to ferment glycerol and reduce nitrate. However, this paradigm was rejected and replaced in the mid-to-late 2000s by genomic analysis that identified at the genetic and molecular levels multiple strains of *Yersinia pestis* originating among host reservoirs (Achtman *et al.* 2004; Drancourt *et al.* 2004; Bos *et al.* 2011; Varlik 2008). Host reservoirs include members of the Cricetidae family of rodents including gerbils, voles, lemmings, and New World rats and mice (Gage and Kosoy 2005), and in Central Asia, marmots (Suntsov and Suntsova 2008).

What does the future hold?

Today, melodramatic reporting in the United States, Asia and Europe of emergent diseases is 'epidemic'. Both the swine flu and Ebola scares in the United States serve to illustrate the political, racial, and socio-economic divisions that pervade the USA, as the fear of travelers and migrants reached outrageous portions! It took comedians like Jon Stewart and Trevor Noah to illustrate the hypocrisy of both the United States mainstream media and many United States politicians during the recent Ebola scare in 'Spot the Africa' (*Daily Show* December 4, 2014). In their sketch, Trevor Noah from South Africa admits he hesitated to visit a country with Ebola:

Trevor Noah "I'm still a bit nervous to be honest, between your cops and frankly your Ebola."

Jon Stewart "Haha – your Ebola, my friend, it's not our, believe me he mis-spoke, you are from Africa, it's your Ebola my friend!"

Trevor Noah "No, no, no – South Africa, John. We haven't had a single case in over 18 years, in fact my friends warned me – they were like, 'Trevor don't go, don't go to the US you'll catch Ebola'. And I was like, you know what guys, just because they had a few cases of Ebola, doesn't mean we should cut travel there. That would be ignorant right!

Jon Stewart "Sure that would be ... that would be ignorant!"

The Daily Show, December 4, 2014

But in spite of the media-generated anxiety, both swine flu and Ebola lacked one of the three optimizing conditions to precipitate a deadly global epidemic: high transmissibility with a relatively high R_0 number, high pathogenicity and lethality, and long-term environmental persistence or frequent reinfection from non-human host reservoirs (Bossak and Welford 2010). In fact, the 2009 swine flu pandemic exhibited high transmissibility but luckily low pathogenicity and lethality. Since March 2014, the Ebola outbreak of West Africa has exhibited relatively low transmissibility (the transmission pathway is blood-to-blood) but high pathogenicity and lethality with an average case fatality of 50% (WHO 2016b). Although Ebola seems to have significant environmental persistence within bats (Leroy *et al.* 2005), the March 2014 epidemic seems to have originated from a cross-species transmission from fruit bats to a single human (Gire *et al.* 2014). Many commentators believe the global human race dodged a bullet with Ebola (e.g., Evans 2015); however, the approximately 11,300 people who died in the West African Ebola epidemic should not be forgotten, nor their lives and deaths minimized by such rhetoric of fortunate escape. Their deaths should instead serve as a warning to us all; somewhere, a virus or bacteria is lurking, just like plague was in 541 CE and 1346 CE, optimized for high transmissibility, high pathogenicity and lethality, and long-term environmental persistence. This is particularly true today. We live in an increasingly smaller world where millions of people move each day round the globe with comparative ease and frequency and could act, at any time, as perfect vectors for a highly lethal, transmissible, newly emergent microorganism (Watts and Strogatz 1998; Bossak and Welford 2010).

Our relationship with the natural world has proceeded through four great periods, and each has seen the emergence of new diseases (McMichael 2004). The first period occurred 5,000–10,000 years ago when humans first became sedentary, living in permanent settlements, cultivating the land and domesticating animals. There is strong evidence during this period indicating a domesticated animal origin for measles and pertussis; however, human modification of the environment might explain other crossovers such as tuberculosis (Pearce-Duvet 2006). The second period occurred 1,000–3,000 years ago when we began to trade across continents (Watts and Strogatz 1998; Bossak and Welford 2010), for instance, the Silk Route and the Greek traders who followed Alexander the Great into India. The Silk Route might have been the conduit for the beginning of the second plague pandemic, the primary wave of the medieval Black Death of 1346–1353. A third period began in 1492 as humans became intercontinental travelers, and in this case European settlers brought Native Americans measles, flu and smallpox. For decades, it was thought that Native Americans gave Europeans syphilis, but four skeletons from the 14th century in Hull, England (Roberts *et al.* 2013) and evidence from victims of Vesuvius found in Pompeii (Armelagos *et al.* 2012) suggest syphilis was already well-established in Europe before Columbus set sail. Today, we live in the fourth great transition, as our global transportation networks move goods and people and diseases, e.g., SARS, rapidly around the globe. Simply put, nearly every place on earth is just 48 hours away from every other place.

Today the changes in land use; changes in human demographics (e.g., increases in population size, densities and urbanization); high malnutrition levels and high prevalence of HIV; the failure of health programs and the emergence of hospital-acquired infections; increased pathogenic virulence and drug resistance; contaminated water supplies; widespread and rapid international travel and trade; and climate change, are primed to facilitate the emergence of new diseases (Woolhouse and Gowtage-Sequeria 2005). Add to these factors persistent poverty (especially peri-urban slums); the increased number and movement of political, economic and environmental refugees; conflict and warfare; and 'superspreaders', all of which increase the likelihood of future pandemics (Bossak and Welford 2010; Weiss and McMichael 2004). For example, persons with AIDS can and have acted as superspreaders of tuberculosis (Weiss and McMichael 2004), while two to three people were responsible for the global spread of SARS (Shen *et al.* 2004). The mystery of the next great pandemic is not *if*, but rather *when* and *where* it will emerge. Therefore, a thorough understanding of the greatest pandemics to affect humankind in the past, those attributed to plague, is imperative.

What does this book address?

The questions that this book will address are: what factors contributed to the emergence, spread, and persistence of plague, and what have been the spatial-temporal impacts of plague on human demography, politics, culture, and economics across recent human history, beginning with the Greek civilization? Although we have evidence that plague evolved from *Y. pseudotuberculosis*, what critical thresholds of human density, human connectivity, and human health allow plague to persist? More broadly, what genetic and biological factors, physical environmental factors, ecological factors, and socio, political and economic factors facilitated plague's emergence from non-human reservoirs and optimized plague's infectivity, lethality, and environmental persistence to cause some of the deadliest pandemics across people, cultures, and history?

The book will be divided into the First, Second, and Third Pandemics. The First Pandemic will include Chapter 2, evaluating the likelihood that plague caused the Antoine Plague, and Chapter 3, reviewing the Justinianic Plague. The Second Pandemic includes those plague pandemics discussed in Chapter 4, the primary wave of the medieval Black Death from 1346–1353, and Chapter 5, those pandemics occurring from 1361–1879. Most of the discussions will concentrate on Europe because of a lack of data from other regions, although the plague was not restricted to Europe from 1346 to 1879. Thereafter, plagues of the Third Pandemic, which began in Hong Kong in 1894 (Figure 1.2), will be discussed in Chapters 6 and 7. The book will conclude with a chapter evaluating the likelihood that plague could be used as a biological weapon, and considering modern attempts to control, monitor, and model plague outbreaks.

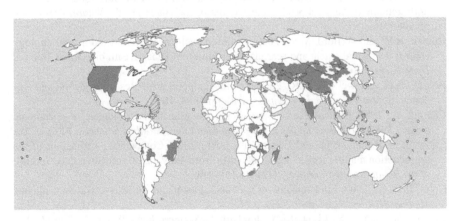

▓▓ Areas* with potential plague natural foci based
on historical data and current information

* First administrative level representation

Figure 1.2 Global distribution of the Third Plague Pandemic as of March 2016.
(Reprinted with permission from WHO/PED (WHO 2016a), as of
March 15, 2016.)

References

Achtman, M., Morelli, G., Zhu, P., Wirth, T., Diehl, I., Kusecek, B., Vogler, A.J., Wagner, D.M., Allender, C.J. and Easterday, W.R., 2004. Microevolution and history of the plague bacillus, Yersinia pestis. *Proceedings of the National Academy of Sciences of the United States of America*, *101*, pp.17837–17842.

Achtman, M., Zurth, K., Morelli, G., Torrea, G., Guiyoule, A. and Carniel, E., 1999. Yersinia pestis, the cause of plague, is a recently emerged clone of Yersinia pseudotuberculosis. *Proceedings of the National Academy of Sciences*, *96*(24), pp.14043–14048.

Ari, T.B., Neerinckx, S., Gage, K.L., Kreppel, K., Laudisoit, A., Leirs, H. and Stenseth, N.C., 2011. Plague and climate: scales matter. *PLoS Pathogens*, *7*(9), p.e1002160.

Armelagos, G.J., Zuckerman, M.K. and Harper, K.N., 2012. The science behind pre-columbian evidence of syphilis in Europe: research by documentary. *Evolutionary Anthropology*, *21*(2), pp.50–57.

Begier, E.M., Asiki, G., Anywaine, Z., Yockey, B., Schriefer, M.E., Aleti, P., Ogen-Odoi, A., Staples, J.E., Sexton, C., Bearden, S.W. and Kool, J.L., 2006. Pneumonic plague cluster, Uganda, 2004. *Emerging Infectious Diseases*, *12*(3), p.460.

Bellazzini, M.A. and Myint, M., 2006. Yersinia Pestis: new challenges in the age of bioterrorism with self-assessment test. *Resident and Staff Physician*, *52*(3), p.21.

Biggerstaff, M., Cauchemez, S., Reed, C., Gambhir, M. and Finelli, L., 2014. Estimates of the reproduction number for seasonal, pandemic, and zoonotic influenza: a systematic review of the literature. *BMC Infectious Diseases*, *14*(1), p.1.

Bos, K.I., Herbig, A., Sahl, J., Waglechner, N., Fourment, M., Forrest, S.A., Klunk, J., Schuenemann, V.J., Poinar, D., Kuch, M. and Golding, G.B., 2016. Eighteenth century Yersinia pestis genomes reveal the long-term persistence of an historical plague focus. *Elife*, *5*, p.e12994.

Bos, K.I., Schuenemann, V.J., Golding, G.B., Burbano, H.A., Waglechner, N., Coombes, B.K., McPhee, J.B., DeWitte, S.N., Meyer, M., Schmedes, S. and Wood, J., 2011. A draft genome of Yersinia pestis from victims of the Black Death. *Nature*, *478*(7370), pp.506–510.

Bossak, B.H. and Welford, M.R., 2009. Did medieval trade activity and a viral etiology control the spatial extent and seasonal distribution of Black Death mortality? *Medical Hypotheses*, *72*(6), pp.749–752.

Bossak, B.H. and Welford, M.R., 2010. Spatio-Temporal attributes of pandemic and epidemic diseases. *Geography Compass*, *4*(8), pp.1084–1096.

Chernin, E., 1989. Richard Pearson Strong and the Manchurian epidemic of pneumonic plague, 1910–1911. *Journal of the History of Medicine and Allied Sciences*, *44*(3), p.296.

Chowell, G., Nishiura, H. and Bettencourt, L.M., 2007. Comparative estimation of the reproduction number for pandemic influenza from daily case notification data. *Journal of the Royal Society Interface*, *4*(12), pp.155–166.

Cohn, Jr., S.K., 2008. Epidemiology of the Black Death and successive waves of plague. *Medical History. Supplement*, (27), p.74.

Cui, Y., Yu, C., Yan, Y., Li, D., Li, Y., Jombart, T., Weinert, L.A., Wang, Z., Guo, Z., Xu, L. and Zhang, Y., 2013. Historical variations in mutation rate in an epidemic pathogen, Yersinia pestis. *Proceedings of the National Academy of Sciences*, *110*(2), pp.577–582.

Daily Show December 4, 2014. www.youtube.com/watch?v=AHO1a1kvZGo, downloaded August 16, 2016.

Devignat, R., 1951. Varietes de l'espece Pasteurella pestis: nouvelle hypothese. *Bulletin of the World Health Organization*, *4*(2), p.247.

Drancourt, M., Roux, V., Tran-Hung, L., Castex, D., Chenal-Francisque, V., Ogata, H., Fournier, P.E., Crubézy, E. and Raoult, D., 2004. Genotyping, Orientalis-like Yersinia pestis, and plague pandemics. *Emerging Infectious Diseases*, *10*(9), pp.1585–1592.

Evans, G., 2015. Dodging the Ebola Bullet. www.ahcmedia.com/blogs/2-hicprevent/post/136658-too-many-hospitals-gambled-on-ebola-preparedness-leaving-them-less-prepared-for-the-next-pandemic, accessed August 16, 2016.

Gage, K.L. and Kosoy, M.Y., 2005. Natural history of plague: perspectives from more than a century of research. *Annual Review of Entomology*, *50*, pp.505–528.

Gani, R. and Leach, S., 2004. Epidemiologic determinants for modeling pneumonic plague outbreaks. *Emerging Infectious Diseases*, *10*(4), p.608.

Gire, S.K., Goba, A., Andersen, K.G., Sealfon, R.S., Park, D.J., Kanneh, L., Jalloh, S., Momoh, M., Fullah, M., Dudas, G. and Wohl, S., 2014. Genomic surveillance elucidates Ebola virus origin and transmission during the 2014 outbreak. *Science*, *345*(6202), pp.1369–1372.

Inglesby, T.V., Dennis, D.T., Henderson, D.A., Bartlett, J.G., Ascher, M.S., Eitzen, E., Fine, A.D., Friedlander, A.M., Hauer, J., Koerner, J.F. and Layton, M., 2000. Plague as a biological weapon: medical and public health management. *JAMA*, *283*(17), pp.2281–2290.

Johnson, N.P. and Mueller, J., 2002. Updating the accounts: global mortality of the 1918–1920 "Spanish" influenza pandemic. *Bulletin of the History of Medicine*, *76*(1), pp.105–115.

Kool, J.L. and Weinstein, R.A., 2005. Risk of person-to-person transmission of pneumonic plague. *Clinical Infectious Diseases*, *40*(8), pp.1166–1172.

Leroy, E.M., Kumulungui, B., Pourrut, X., Rouquet, P., Hassanin, A., Yaba, P., Délicat, A., Paweska, J.T., Gonzalez, J.P. and Swanepoel, R., 2005. Fruit bats as reservoirs of Ebola virus. *Nature*, *438*(7068), pp.575–576.

McMichael, A.J., 2004. Environmental and social influences on emerging infectious diseases: past, present and future. *Philosophical Transactions of the Royal Society of London B: Biological Sciences*, *359*(1447), pp.1049–1058.

Nishiura, H., Schwehm, M., Kakehashi, M. and Eichner, M., 2006. Transmission potential of primary pneumonic plague: time inhomogeneous evaluation based on historical documents of the transmission network. *Journal of Epidemiology & Community Health*, *60*(7), pp.640–645.

Pearce-Duvet, J., 2006. The origin of human pathogens: evaluating the role of agriculture and domestic animals in the evolution of human disease. *Biological Reviews*, *81*(3), pp.369–382.

Perry, R.D. and Fetherston, J.D., 1997. Yersinia pestis – etiologic agent of plague. *Clinical Microbiology Reviews*, *10*(1), pp.35–66.

Rasmussen, S., Allentoft, M.E., Nielsen, K., Orlando, L., Sikora, M., Sjögren, K.G., Pedersen, A.G., Schubert, M., Van Dam, A., Kapel, C.M.O. and Nielsen, H.B., 2015. Early divergent strains of Yersinia pestis in Eurasia 5,000 years ago. *Cell*, *163*(3), pp.571–582.

Ratovonjato, J., Rajerison, M., Rahelinirina, S. and Boyer, S., 2014. Yersinia pestis in Pulex irritans fleas during plague outbreak, Madagascar. *Emerging Infectious Diseases*, *20*(8), p.1414.

Roberts, C.A., Millard, A.R., Nowell, G.M., Gröcke, D.R., Macpherson, C.G., Pearson, D.G. and Evans, D.H., 2013. Isotopic tracing of the impact of mobility on infectious disease: the origin of people with treponematosis buried in Hull, England, in the late medieval period. *American Journal of Physical Anthropology*, *150*(2), pp.273–285.

Schmid, B.V., Büntgen, U., Easterday, W.R., Ginzler, C., Walløe, L., Bramanti, B. and Stenseth, N.C., 2015. Climate-driven introduction of the Black Death and successive plague reintroductions into Europe. *Proceedings of the National Academy of Sciences* 112(10), pp. 3020–3025.

Seal, S.C., 1960. Epidemiological studies of plague in India: 1. The present position. *Bulletin of the World Health Organization, 23*(2–3), p.283.

Shen, Z., Ning, F., Zhou, W., He, X., Lin, C., Chin, D.P., Zhu, Z. and Schuchat, A., 2004. Superspreading SARS events, Beijing, 2003. *Emerging Infectious Diseases, 10*(2), pp.256–260.

Simonet, M., Riot, B., Fortineau, N. and Berche, P., 1996. Invasin production by Yersinia pestis is abolished by insertion of an IS200-like element within the inv gene. *Infection and Immunity, 64*(1), pp.375–379.

Smith, G.J., Vijaykrishna, D., Bahl, J., Lycett, S.J., Worobey, M., Pybus, O.G., Ma, S.K., Cheung, C.L., Raghwani, J., Bhatt, S. and Peiris, J.M., 2009. Origins and evolutionary genomics of the 2009 swine-origin H1N1 influenza A epidemic. *Nature, 459*(7250), pp.1122–1125.

Spyrou, M.A., Tukhbatova, R.I., Feldman, M., Drath, J., Kacki, S., de Heredia, J.B., Arnold, S., Sitdikov, A.G., Castex, D., Wahl, J. and Gazimzyanov, I.R., 2016. Historical Y. pestis genomes reveal the European Black Death as the source of ancient and modern plague pandemics. *Cell Host & Microbe, 19*(6), pp.874–881.

Suntsov, V.V. and Suntsova, N.I., 2008. Concepts of macro-and microevolution as related to the problem of origin and global expansion of the plague pathogen Yersinia pestis. *Biology Bulletin, 35*(4), pp.333–338.

Sussman, G.D., 2011. Was the Black Death in India and China? *Bulletin of the History of Medicine, 85*(3), pp.319–355.

Varlik, N., 2008. *Disease and empire: a history of plague epidemics in the early modern Ottoman Empire (1453–1600).* Chicago: The University of Chicago.

Wagner, D.M., Klunk, J., Harbeck, M., Devault, A., Waglechner, N., Sahl, J.W., Enk, J., Birdsell, D.N., Kuch, M., Lumibao, C. and Poinar, D., 2014. Yersinia pestis and the plague of Justinian 541–543 AD: a genomic analysis. *The Lancet Infectious Diseases, 14*(4), pp.319–326.

Watts, D.J. and Strogatz, S.H., 1998. Collective dynamics of "small-world" networks. *Nature, 393*(6684), p.440.

Weiss, R.A. and McMichael, A.J., 2004. Social and environmental risk factors in the emergence of infectious diseases. *Nature Medicine, 10*(12s), p.S70.

Wheelis, M., 2002. Biological warfare at the 1346 Siege of Caffa. *Emerging Infectious Diseases,* 8 (9), pp. 971–975.

Woolhouse, M.E. and Gowtage-Sequeria, S., 2005. Host range and emerging and reemerging pathogens. *Emerging Infectious Diseases, 11*(12), p.1842.

World Health Organization, 2016a. Global distribution of natural plague as of March 2016. Reprinted from WHO/PED, as of March 15, 2016. www.who.int/csr/disease/plague/Plague-map-2016.pdf?ua=1

World Health Organization, 2016b. Ebola virus disease. www.who.int/mediacentre/factsheets/fs103/en/, downloaded August 16, 2016.

2 The Athenian Pandemic

One of the first known pandemics to ravage Europe arrived in Athens in 432 BCE. It struck Athens at a time before Alexander the Great had invaded western India in 326 BCE or the overland Silk Road had commenced in the 2nd century BCE. Although westerners consider the Greek Empire a large and highly integrated empire, before Alexander the Great it was a relatively small, disconnected entity that knew little about the vast, wealthy civilizations of China and India. In other words, the world consisted of civilizations disconnected from each other, where local epidemics raged but rarely, if ever, spawned global pandemics. In disease terms, disconnected populations favor the evolution of lower transmissibility and lower virulence because the parasite or pathogen will likely infect all possible hosts, killing them and becoming extinct before escaping the confines of a region (Boots and Mealor 2007). In other words, diseases were endemic and restricted to specific regions at the time of the Athenian Pandemic.

In 431 BCE Athens was engaged in the Peloponnesian War with Sparta. The next year, plague struck Athens. The scale and widespread physical and moral devastation of this plague was greater in scope than the Peloponnesian War; in fact, tens of thousands died of plague (Luginbill and Ferrill 1999). The 50 years prior to the Peloponnesian War were marked by the consolidation and expansion of the initial Delian League that became the Athenian Empire. By 431 BC, this Empire controlled much of the coastline and islands of the Aegean Sea, as well as the Dardanelles and Bosphorus (collectively called the Turkish Straits) that separate Europe from Asia. (In fact, Asia (Ἀσία) is the Greek word for the east bank of the Aegean Sea). Also by 431 BC, the Athenian Empire had become the intermediary between Europe and Asia (Wood 1993). As a result of its strategic position, the Athenian Empire became a conduit for new scientific and philosophical ideas that emerged from Iraq, Iran, and India during the Axis Age (Wood 1993), new technology and trade goods, and without a doubt new zoonotic diseases (those diseases transmitted from animals to humans). This convergence of ideas and trade led to one of the great eras in world culture – the Hellenistic Age (Wood 1993) and one of the earliest forms of archaic globalization linking western Europe, the Middle East, and Far East. If we can accept Thucydides' 450 BCE estimates that around 1.1 million people were living in the Peloponnesian League, 2.5 million in the Athenian Empire and between 0.7–1.8 million in the Persian Empire (Ober 2015), then pandemic threshold densities (N_T) were favorable for emergent pandemic diseases. Certainly, increasing human population densities and increasing movement of people due to war and economic collapse increase the risk of new emergent diseases (Wolfe *et al.* 2005). Similarly, in Africa, the early emergence of HIV/AIDS in the 20th century was facilitated by increased rural-urban migration and higher between-person contact rates (Wolfe *et al.* 2005). Later the global emergence of HIV/AIDS followed the expansion of road networks in central Africa, an increase in the proportion of migrating males, and an increase in international travel (Wolfe *et al.* 2005).

It would appear that by 431 BCE the Athenian Empire was large enough and connected enough to animal sources of disease to initiate, transmit, and sustain the first known 'global' pandemic disease. The Peloponnesian War with Sparta would appear to have increased rural-urban migration and between-person contacts, weakened immune systems as the Athenian Empire struggled to provide necessary human services such as food and waste disposal, and provided the means to disperse the disease across the Athenian Empire.

Arrival

Originating possibly in Ethiopia, the Athenian Pandemic spread to Egypt and Libya and the Persian Empire (Morens and Littman 1992; Lattimore 1998; Retief and Cilliers 1998), before this plague reached Athens and Lemnos via the harbor of Piraeus. In fact, the Athenian Pandemic hit Athens shortly after the arrival of the Lacedaimonian army in Attica (Longrigg 1980). Thereafter the Athenian Pandemic made its way up into the hills surrounding Athens, where it would continue to linger and reappear periodically from the summer of 430 BCE until 427 or 426 BCE (Luginbill and Ferrill 1999; Morens and Littman 1992; Retief and Cilliers 1998; Tritle 2004). This suggests the disease was transmitted to Athens via trade or naval activity. According to Thucydides, the Peloponnesian peninsula was spared, as were the besieging Spartan soldiers (Retief and Cilliers 1998). Thucydides notes that more than 3,000 Athenians of Hagnon's July 430 BCE naval expedition to besiege Potidaea died of the Athenian Pandemic over the course of six weeks (Morens and Littman 1992). Whereas, Thucydides mentions that the besieged Potidaeans died of starvation or cannibalism rather than the Athenian Pandemic (Morens and Littman 1992). Elsewhere Livy (a Roman historian who lived from 59 BCE to 17 CE) identifies plague in Rome in 433 and 428 BCE (Morens and Littman 1992).

Approximately a third of the Athenian population perished, contributing to one of the major reverses suffered by the Athenians during the Peloponnesian War (Luginbill and Ferrill 1999; Lattimore 1998). As bodies were burned on flaming pyres in the streets, the universal feeling was that it was best to live for the moment, as life as one knew it was coming to an end (Luginbill and Ferrill 1999). The Athenian Pandemic knew no boundaries nor distinctions between young and old, rich and poor (Luginbill and Ferrill 1999). Lawlessness prevailed; the rich basked in pleasure as they attempted to squander their fortunes, trying to prevent the poor from profiting from their apparently inevitable deaths (Luginbill and Ferrill 1999).

A mass grave dated to between 430 and 426 BCE, found adjacent to Athens' ancient Kerameikos cemetery, points to a state of mass panic (Baziotopoulou-Valavani 2002). The few grave-offerings were mostly common, cheap burial vessels, while most of the 150 skeletons lay helter-skelter, forming more than five layers, with no soil placed between the layers of bodies, and all were apparently placed there within a day or two (Baziotopoulou-Valavani 2002; Papagrigorakis *et al.* 2006). The lower levels of bodies were spaced out, suggesting some care was taken in their burial (Papagrigorakis *et al.* 2006). However, the upper layers of bodies appear to be thrown into the pit in a chaotic manner (Papagrigorakis *et al.* 2006). The age of burial, lack of burial vessels, helter-skelter positions of the bodies, and lack of soil between the skeletons all point to a plague burial, not the burial of war dead, which would be more orderly (Baziotopoulou-Valavani 2002; Papagrigorakis *et al.* 2006). It is suspected that this mass grave was filled with the poor whose surviving family members neither had the time, inclination, nor money for proper burials (Baziotopoulou-Valavani 2002).

Symptoms

Sickness first appeared in the form of fever, accompanied by a feeling of hopelessness, that progressed to congestion, vomiting, and convulsions. Thucydides' eyewitness account noted the abrupt onset of fever, headache, fatigue and stomach pain followed by vomiting and – if a sufferer survived this early onslaught – after seven days, severe diarrhea (Thucydides 1900; Kazanjian 2015). Bleeding from the mouth also occurred, while fatal dehydration frequently followed diarrhea (Thucydides 1900; Kazanjian 2015). Fear of contracting the disease prevented many people from caring for the sick; those who did so out of honor were often stricken with plague (Luginbill and Ferrill 1999; Davidson 2011). Thucydides noted that among the first victims of the Athenian Pandemic were the physicians and caregivers (Thucydides 1900; Kazanjian 2015). Again, according to Thucydides, the disease indiscriminately attacked all manner of people, but did not reinfect those who had survived early attacks, nor was the disease restricted to conditions of severe overcrowding, as Thucydides noted the plague returned to Athens in 427/426 BCE when it was not being besieged (Morens and Littman 1992).

The sheer magnitude of this disastrous event caused people to stop mourning for their own relatives, and a feeling of euphoria was experienced by survivors who thought themselves immune (Luginbill and Ferrill 1999). Certainly, the Athenian Pandemic overwhelmed any attempt at public health in Athens, and the capabilities of the medical practitioners of the time; no remedies were demonstrably successful (Hays 2005). However, Galen's legend of Hippocrates' use of fire to combat the Athenian Pandemic (the idea being that 'bad' air caused disease and that the controlled use of fires prevented the spread of disease) survived well past the Renaissance (Pinault 1986). Such notions seriously impeded modern understandings of pathogenic disease, which had to wait until the mid-1800s, with Snow and his ideas connecting cholera to an infectious agent.

Speed, seasonality, duration, and mortality of the Athenian Pandemic

Data on the Athenian Pandemic are vague at best, and this is primarily because Thucydides' account of the disease is recounted in passing while he concentrates on the Peloponnesian War, and therefore the temporal component leaves much to be desired (Davidson 2011). In fact, the only cities with documented epidemic initiation and termination dates are Piraeus and Athens. No dates are given for Ethiopia, Egypt, Libya, Persia or Lemnos, although these areas are mentioned in the chronology of the Athenian Pandemic. Thucydides merely chronicles where the disease was, from whence it came, and where it went (Davidson 2011). However, it is generally agreed that the Athenian Pandemic originated in Ethiopia, and then traveled through Egypt and Libya before reaching the territory of the king of Persia, and finally reaching Athens by way of the Piraeus harbor below the city (Bray 1996; McNeill 1976; Lattimore 1998). The Athenian Pandemic struck both Piraeus and Athens in the summer of 430 BCE and lasted intermittently until the summer of 428 (Retief and Cilliers 1998; Luginbill and Ferrill 1999; Davidson 2011).

The first wave of the Athenian Pandemic occurred in early May 430 BCE, soon after the Spartan army laid siege to Athens. Thereafter, pandemics hit in summer 428 BCE and again in the 427–426 BCE winter (Morens and Littman 1992). The population of Athens might have swollen to as much as 400,000 in early May 430 BCE as refugees entered the city ahead of the Spartan army, although Athens' base population was closer to 155,000 (Rostovcev 1941; Morens and Littman 1992). Thucydides suggests the epidemic lasted 4–5 years with a secondary peak in summer 428 BCE (Morens and Littman 1992).

Thucydides says that between July–August 430 BCE 26% of a 4,000-strong Hoplite expedition died in 40 days (Morens and Littman 1992). Thucydides himself contracted the Athenian Pandemic but survived, as did many others, and he appears not to have been infected again (Morens and Littman 1992). All told, case fatalities of 25% have been suggested for the duration of this non-seasonal, 4-year-long epidemic (Morens and Littman 1992). Among the notable casualties was Pericles, who died in the second year of the Athenian Pandemic. In losing Pericles, Athens lost a great statesman, orator, and soldier, though as a soldier he might have increased the fatality rate within Athens by forcing tens of thousands of people into Athens as Sparta surrounded Athens. It is believed that as a result of Pericles' actions between 250,000 and 400,000 people were living in squalid conditions within the 4-square-mile area of Athens when the Athenian plague struck (Durack *et al.* 2000). The resulting unsanitary conditions would create ideal conditions for mosquito, flea, tick, louse, human-human, or water-borne pathogens to multiply and kill tens of thousands of people.

To summarize, the Athenian Pandemic was probably density-dependent as it affected the citizens of Athens but not the besieging Spartan army. Unlike subsequent plagues caused by *Y. pestis*, the Athenian Pandemic did not have a seasonal signature; it hit in both summer and winter, and it didn't persist for very long; Thucydides suggested it only lasted 4 years (Retief and Cilliers 1998). In contrast, the primary wave of the 1346–1351 medieval Black Death was dominantly a disease of the late summer and early fall and persisted until the mid-1800s and, unlike the Athenian Pandemic, did confer immunity to those that survived.

Social/cultural/economic/political ramifications/environmental change

The fear and panic precipitated by the Athenian Pandemic undermined Greek civic society. According to Thucydides, Greeks disregarded civic authority, violated laws and customs, discarded burial procedures with thousands of untended corpses being left on the streets of Athens, ignored infected individuals and families, rendering them isolated at the time of their greatest need, and abandoned their work both in the fields and cities (Kazanjian 2015). "The Disease brought the beginning of great lawlessness" and "no fear of god or the law of man restrained them", Thucydides *History* 2.53.1 (Morgan 1994).

The resulting famine co-precipitated by Pericles, the Athenian General and the Spartans isolating Athens behind its defenses, sealed Athens' fate. Moreover, according to Thucydides, the Spartans destroyed abandoned fields and orchards outside Athens (Morgan 1994). "Such was the disaster which fell upon the Athenians crushing them, with people dying inside the city and the land outside laid waste", Thucydides *History* 2.54.1 (Morgan 1994).

It should be emphasized that a major shortcoming of commentaries on the Athenian Pandemic is that they all rely on essentially one primary source: Thucydides. Although an eyewitness, much of Thucydides' *History* discusses the Peloponnesian War although he spends considerable effort describing the symptoms of the Athenian Pandemic (Thucydides *History* 2.54-57, Morgan 1994). Nevertheless, a good argument can be made that together the Peloponnesian War and the Athenian Pandemic contributed to the end of the Golden Age of Athens (Papagrigorakis *et al.* 2006; Kazanjian 2015). All told, 'the plague' killed Pericles and ~25–30% of Athens' population, and both had an immediate effect on the outcome of the Peloponnesian War. Athens lost its leading statesman and highest ranked general, and those who replaced him were less experienced and hence less capable (Littman comment within Durack *et al.* 2000). In killing 20–25% of Athens' population, the Athenian Pandemic severely compromised its fighting capability. Although the Greek army of Athens was well-trained and had a large navy, the Spartan army was also very capable, and, as a result, the under-manned, immune-compromised, poorly led Athenian army (in a post-Pericles environment) could not match or defeat Sparta (Littman's comment within Durack *et al.* 2000).

The fear and panic that affected Athens during the Athenian Pandemic was also observed by Boccaccio in Italy during the medieval Black Death and during the early period of the AIDS epidemic, and it was certainly observed during the recent Ebola epidemic in West Africa (Kazanjian 2015).

Etiologic agent

Although Thucydides provided a detailed description of the 430 BCE disease, much of this from his own exposure to the disease, conclusive proof of etiology has not been forthcoming until recently. Several pathogens have been implicated: Ebola (Kazanjian 2015), typhoid fever and epidemic typhus, anthrax, smallpox (Littman and Littman 1969; Retief and Cilliers 1998), plague and flu (Papagrigorakis *et al.* 2006). DNA from three teeth from 150 skeletons excavated from Kerameikos ancient cemetery of Athens and dated to 430 BCE were tested using 'suicide' polymerase chain reaction (PCR) amplification protocols (Papagrigorakis *et al.* 2006). Six putative causal agents of the Athenian Pandemic were tested separately. These were plague, typhus, anthrax, tuberculosis, cowpox and cat-scratch disease, and none yielded any 'suicide' reaction product (Papagrigorakis *et al.* 2006). However, the ancestral strain of *Salmonella enterica* did yield PCR reaction product, indicating the presence in the skeletal teeth of typhoid (Papagrigorakis *et al.* 2006). What's more, the strain of *Salmonella enterica* serovar Typhi located in the teeth at the mass burial found adjacent to the Kerameikos ancient cemetery of Athens might be the original strain of *Salmonella enterica* that today is exclusively adapted to humans (Papagrigorakis *et al.* 2007). Thucydides' report that animals were also affected by this disease supports this contention (Papagrigorakis *et al.* 2007). Although widely accepted today, Shapiro *et al.* (2006) did contest this analysis based on the possibility of soil contamination of one gene sequence. If the ancestral strain of *Salmonella enterica* was responsible for the Athenian plague, this could explain its high 33% mortality (Luginbill and Ferrill 1999; Lattimore 1998). It is also generally accepted that plague (*Y. pestis*) evolved in Central Asia (Gage and Kosoy 2005), which during the 5th century BCE was disconnected from Greece and thus is an unlikely candidate for the Athenian Pandemic.

What we can say conclusively about the Athenian Pandemic is that it illustrates the beginning, however fragile, of the connection of Europe with the Middle East and Central Asia: the creation of an integrated Eurasian culture, one that was just beginning to share technology, culture, and disease. The Athenian Pandemic is one product of this 'proto-globalization' process.

References

Baziotopoulou-Valavani, E., 2002. A mass burial from the cemetery of Kerameikos. *Excavating Classical Culture. Recent Archaeological Discoveries in Greece. Studies in Classical Archaeology I. BAR International Series, 1031*, pp.187–201.
Boots, M. and Mealor, M., 2007. Local interactions select for lower pathogen infectivity. *Science, 315*(5816), pp.1284–1286.

Bray, R.S., 1996. *Armies of pestilence: the effects of pandemics on history*. Cambridge: Lutterworth Press.

Davidson, A.L., 2011. *Disputed causality of the Justinianic and Athenian plagues: analysis through spatio-temporal mapping*. An unpublished bachelor's thesis, Georgia, USA: Georgia Southern University.

Durack, D.T., Littman, R.J., Benitez, R.M. and Mackowiak, P.A., 2000. Hellenic holocaust: a historical clinico-pathologic conference. *The American Journal of Medicine*, *109*(5), pp.391–397.

Gage, K.L. and Kosoy, M.Y., 2005. Natural history of plague: perspectives from more than a century of research. *Annual Review of Entomology*, *50*, pp.505–528.

Hays, J.N., 2005. *Epidemics and pandemics: their impacts on human history*. ABC-CLIO.

Kazanjian, P., 2015. Ebola in antiquity? *Clinical Infectious Diseases*, *61*(6), pp.963–968.

Lattimore, S., 1998. *The Peloponnesian war*. Indianapolis: Hackett Publishing.

Littman, R.J. and Littman, M.L., 1969. The Athenian Plague: smallpox. *Transactions and Proceedings of the American Philological Association*, *100*, pp.261–275.

Longrigg, J., 1980. The great plague of Athens. *History of Science*, *18*(3), pp.209–225.

Luginbill, R.D. and Ferrill, A., 1999. *Thucydides on war and national character*. Oxford: Westview Press.

McNeill, W.H., 1976. *Plagues and peoples*. New York: Anchor.

Morens, D.M. and Littman, D.M., 1992. Epidemiology of the plague of Athens. *Transactions of the American Philological Association*, *122*, pp.271–304.

Morgan, T.E., 1994. Plague or poetry? Thucydides on the epidemic at Athens. *Transactions of the American Philological Association (1974–)*, *124*, pp.197–209.

Ober, J., 2015. *The rise and fall of classical Greece*. New York: Princeton University Press.

Papagrigorakis, M.J., Synodinos, P.N. and Yapijakis, C., 2007. Ancient typhoid epidemic reveals possible ancestral strain of Salmonella enterica serovar Typhi. *Infection, Genetics and Evolution*, *7*(1), pp.126–127.

Papagrigorakis, M.J., Yapijakis, C., Synodinos, P.N. and Baziotopoulou-Valavani, E., 2006. DNA examination of ancient dental pulp incriminates typhoid fever as a probable cause of the plague of Athens. *International Journal of Infectious Diseases*, *10*(3), pp.206–214.

Pinault, J.R., 1986. How Hippocrates cured the plague. *Journal of the History of Medicine and Allied Sciences*, *41*(1), pp.52–75.

Retief, F.P. and Cilliers, L., 1998. The epidemic of Athens, 430–426 BC. *South African Medical Journal*, *88*(1), pp.50–53.

Rostovcev, M.I., 1941. *The social and economic history of the Hellenistic world*. Oxford: Clarendon Press.

Shapiro, B., Rambaut, A. and Gilbert, M.T.P., 2006. No proof that typhoid caused the plague of Athens (a reply to Papagrigorakis et al.). *International Journal of Infectious Diseases*, *10*(4), pp.334–335.

Thucydides, 1900. *History of the Peloponnesian War*. Book II, 137–140, trans. B. Jowett. Oxford: Clarendon Press.

Tritle, L.A., 2004. *The Peloponnesian War*. Westport, USA: Greenwood Publishing Group.

Wolfe, N.D., Daszak, P., Kilpatrick, A.M. and Burke, D.S., 2005. Bushmeat hunting, deforestation, and prediction of zoonotic disease. *Emerging Infectious Diseases*, *11*(12), pp.1822–1827.

Wood, M., 1993. *Legacy: a search for the origins of civilization*. London: Network Books.

3   Antonine Pandemic and Justinianic Plague

First Plague Pandemic begins

Prior to the Antonine Pandemic and Justinianic Plague, the Old World crossed a critical geographic or spatial threshold, it moved from a 'large world' to a 'small world' (Watts and Strogatz 1998). This represents the latter part of the second historic transition where continental-wide trade connections were established and maintained among the early cultural hearths (McMichael 2010). In other words, the Old World consisted of largely isolated empires in the Fertile Crescent, the Indo-Ganges plain, and China prior to the 2nd century BCE and the advent of the embryonic overland Silk Road. Diseases within 'large world' clusters (those relatively isolated civilizations) selected for lower transmission and lower virulence; otherwise diseases would become extinct before escaping their isolated clusters (Boots and Sasaki 1999; Galvani 2003; Boots and Mealor 2007). As a result, these isolated clusters obtained both acquired immunity (those infected who survived) and herd immunity (when a large percentage of the population have acquired immunity and those susceptible are too few and far between to sustain a disease outbreak) to their dominant infections but remained highly susceptible to other isolated clusters' dominant diseases. As a result, within a 'large world', lethal, highly transmissible pandemic diseases are rare. But when dominant infections erupt from one cluster to another cluster that lacks immunity, the results are devastating. For instance, we are all familiar with the millions of deaths wrought by measles, small pox and influenza in the New World following their introduction by Spanish conquistadors in 1492.

The opening of the overland Silk Road in the 2nd century BCE and the expansion of the Hellenic Greek empire that began in 323 BCE irreversibly changed the world. A new 'smaller world' began to emerge: one more connected but where dominant diseases or newly emergent diseases could pass along the new overland Silk Road. This certainly seems to be the case with the Justinianic Plague.

The Antonine Pandemic

The Antonine Pandemic of 165–180 CE preceded the Justinianic Plague by 400 years, and its impact, though not its etiologic identity, was similar (Cunha and Cunha 2008). It is possible the Antonine Pandemic erupted during the Parthian

War (162–166 CE) in Mesopotamia and was brought westwards by returning legionnaires (Greenberg 2003), although no actual evidence supports this conjecture. First reported in Nisibis and Smyrna in 165 CE, this pandemic reached Rome in 166 CE whereupon Galen, the distinguished medical practitioner, fled Rome (Duncan-Jones 1996; Galen in Kühn 1821–1833). This pandemic might have reappeared in 189 CE where, according to Dio Cassius (72.14.3-4), 2,000 people died in Rome in a single day (Littman and Littman 1973).

A large unattributed (possible tropical) volcanic eruption of 169 CE is associated with strong decadal post-volcanic cooling (Sigl *et al.* 2015) and might have played a sustaining role in the Antonine Pandemic of 165–180 CE (Harper 2016). Strong decadal cooling triggered by tropical volcanic activity has been associated with agricultural failures and subsequent famines throughout human history (Lamb 2002). Even today, in an era of modern medicine, famines weaken human immune systems to the point where opportunistic disease, such as flu or even the common cold, kill famine sufferers (Bryce *et al.* 2005). Furthermore, the climate of the last decade of the Antonine Pandemic of 165–180 CE was highly variable, with large swings in temperature and rainfall.

The Antonine Pandemic of 165–180 CE and subsequent outbreaks throughout the following century initiated a severe demographic crisis across the entire breadth of the Roman Empire, as mortalities reached 25%, triggering price and wage inflationary shocks (Scheidel 2002). Public grain dole receipts suggest that 60% or more of Alexandria's population succumbed to the pandemic, while 5,000 a day were dying in Rome at the peak of the epidemic (Harper 2016). However, 7–10% mortality is probably a more reliable measure, with the army suffering 13–15% mortalities (Littman and Littman 1973). The Antonine Pandemic killed two emperors, including Marcus Aurelius in 180 CE, and crippled the Roman army, as conscription could not replace manpower losses as eastern tribes contested the rule of Rome on its eastern fringes, and paved the way for the collapse of the Eastern Roman Empire in the post-Justinianic Plague environment (Fears 2004). Christian persecution was also stepped up as Marcus Aurelius and Rome sought to blame someone for the plague, while portrait busts of the rich and powerful showed both anxiety and fear for the first time (Fears 2004). Yet proxy empirical evidence of higher mortalities (e.g., declines in military diplomas, reductions in the volume of year-dated Egyptian documents, census returns) and socio-economic collapse (e.g., reduction in the volume of non-imperial and imperial building dedications) appear either exaggerated or anomalies, offering no substantial support for the catastrophe (Greenberg 2003).

Mass burials from this period are few and far between, but excavations in Gloucester, UK yielded a mass grave containing 91 randomly aligned skeletons that appear to have been placed there in one episode in the late 2nd century CE (Simmonds *et al.* 2008). The chaotic nature of this and other mass burials in Gloucester and the skeletons' low $\delta^{15}N$ values (indicating low meat-protein diets) suggest these dead were lower-class members of Roman society (Cheung *et al.* 2012). Elsewhere, 315 of a possible total of 1,300 individuals were excavated from a mass single-event burial in the catacomb of Saints Peter and Marcellinus in

Rome from the end of the 2nd century CE (Blanchard *et al.* 2007). Few grave goods accompanied the burials but traces of gold, silver and amber suggest a high-status assemblage (Blanchard *et al.* 2007). The höyük or hill-shaped tomb at Oymaağaç-Nerik, Turkey, adjacent to the Roman town of Neoklaudiopolis yielded over 120 individuals buried in multiple mass graves from the late 2nd century (Marklein and Fox 2016). All these mass burials have been tentatively associated with some communal fatal incident, possibly a pandemic disease.

Etiologically speaking, measles, and epidemic typhus have been suggested as the etiologic (disease) agent, but a more plausible candidate is smallpox (Sabbatani and Fiorino 2009). Galen noted the following: exanthema (spots)-covered patients; patients experiencing high-volume bloody black diarrhea; high fever; vomiting; stomach upset and coughing – all systematic of smallpox, not plague (Littman and Littman 1973). However, the repeated outbreaks of the Antonine Pandemic throughout the Roman Empire from 165–180 CE conflate this argument as smallpox confers lifelong immunity if survived. Although unlikely, repeated outbreaks of smallpox might have occurred in previously unaffected areas (Cunha and Cunha 2008). Nevertheless, until genomic materials are processed from epidemic victims from this period, *Y. pestis* (plague) cannot be ruled out as the cause (Wagner *et al.* 2014).

The Justinianic Plague

The timing of the Justinianic Plague, just like the medieval Black Death, was impeccable; it hit the Eastern Roman Empire ruled by Justinian the Great toward the end of a period of intense climate instability from ~250–550 CE (Büntgen *et al.* 2011; Diaz and Trouet 2014). The European Migration Period and the fall of the Roman Empire occurred during or immediately after this period of intense climate volatility. The European Migration Period, which saw widespread migration of Germanic peoples into southern Europe (e.g., Goths, Vandals) and then more eastern tribes into western and southern Europe (e.g., Huns, Slavs), has also been strongly linked with the fall of the Roman Empire (Huntington 1917; Diaz and Trouet 2014). It would appear that persistent droughts in Central Asia, triggered by the intense climate instability of ~250–550 CE, initiated this European Migration Period, with the Huns seeking fresh forage to the west and south of their ancestral region (Diaz and Trouet 2014). The intensity and persistence of hydroclimatic fluctuations between ~250–550 CE in Europe also contributed to widespread famines at a time when cereal cultivation such as wheat and barley was shifted to more marginal lands as more profitable cultivation of olives and vines expanded (Diaz and Trouet 2014). As a result of the increased frequency of famines and the reduction in the production of food staples that occurred through this 300-year period, the vast majority of Europe's population became increasingly vulnerable, deeply impoverished and malnourished, and increasingly susceptible to newly emerging zoonotic and anthroponotic diseases and persistent insect-vectored diseases such as malaria.

In spite of the multiplicity of famines and the deep impoverishment of most Romans, Emperor Justinian's power was at its height in the 530s CE. He overcame and conquered Germanic invaders in the form of the Ostrogoths in Italy and Visigoths in Spain and was well on his way to reuniting the Western and Eastern Roman Empires (Sohysiak 2008). But the 540s CE saw renewed Persian attacks from Syria, increased Ostrogoth resistance in Italy, and Transdanubian aggression in the Balkans that halted the expansion of the Eastern Roman Empire (Sohysiak 2008). To compound the problems the Eastern Roman Empire was facing in the mid-6th century, the Justinianic Plague hit Europe and the Middle East between two of the sixteen most negative tree-ring growth anomalies (536 CE and 543 CE) between 500 BCE and 1000 CE, in other words, between two of the coldest summers (Sigl *et al.* 2015). Large volcanic signals of ash aerosols and sulfur deposition precede both 536 and 543 CE, and cold summers contribute to the coldest decade (536–545 CE) observed between 500 BCE and 1000 CE (Sigl *et al.* 2015).

The 535–536 CE eruption occurred in the Northern Hemisphere, resulting in a 1.6–2.5°C temperature decline below the 30-year average for the 536 CE summer (Sigl *et al.* 2015) or a reduction of 11.3 watts m^{-2} in sunlight (Büntgen *et al.* 2016). Given that average global ground-surface solar radiation receipt is ~1120 W m^{-2}, a reduction of 11.3 W m^{-2} in sunlight does not seem like much. However, over a year, this represents an approximate 1% reduction in global solar energy receipt at the ground surface. Rather surprisingly, multiple Northern Hemisphere signals are present in ice-cores (Sigl *et al.* 2015), although this eruption has been associated with the Ilopango volcano/caldera in El Salvador (Dull *et al.* 2001). This eruption has been linked with the densest and most persistent dry fog ever recorded in Europe and the Middle East between 536–537 CE (Strothers and Rampino 1983).

The sun was dark and its darkness lasted for eighteen months; each day it shone for about four hours, and still this light was only a feeble shadow ... the fruits did not ripen and the wine tasted like sour grapes.

Probably John of Ephesus
(Strothers and Rampino 1983)

The 539–540 CE eruption was tropical in origin, yielding 10% greater aerosol loading than the 1815 Tambora eruption (Sigl *et al.* 2015) or a 19.1 W m^{-2} reduction in sunlight (Büntgen *et al.* 2016). It appears highly likely that a massive eruption, possibly by Rabaul, Indonesia 4°S in 540 CE ±10 years (Strothers and Rampino 1983), was responsible for this event. A smaller eruption that reduced sunlight by 1.1 W m^{-2} occurred in 547 CE (Büntgen *et al.* 2016). Together the three eruptions of 536, 540 and 547 CE transformed the Late Antique world by initiating the Late Antique Little Ice Age that ran from 536 to 660 CE (Büntgen *et al.* 2016). The cold and subsequent drought led to a Mediterranean and Mesopotamian-wide famine, and drought and famine in China between 536 and 538 CE (Weisburd 1985). The famine in China was so severe that 70–80% of the population died (Pang *et al.* 1989). European demographic collapse is evident even in the Mälaren Valley in Sweden (Löwenborg 2012). Tree-ring sequences and stalagmite laminae support this global cooling (Bailhe 1995; Büntgen *et al.* 2011; Baker *et al.* 2015). In fact, tree-ring data indicate that 536 CE was the second coldest summer in the last 1,500 years, while tree-ring sequences from both North America and Europe indicate that climates remained volatile and cold for the next 15 years (Bailhe 1995; Büntgen *et al.* 2011). Recent work suggests a 3°C cooling for more than a decade (Larsen *et al.* 2008; McMichael 2012). It is probably not coincidental that an opportunistic disease, the Justinianic Plague (the world's first known *Yersinia pestis* plague pandemic) erupted across Eurasia at this time among the impoverished, malnourished and health-compromised Eurasians.

Certainly, both animal studies and recent work on the impacts of the Dutch 1944–1945 famine suggest that famines compromise subsequent human health for decades, especially for those children born or conceived immediately before or during a famine (Roseboom *et al.* 2006). Evidence suggests that climate anomalies affect health in three ways: (1) direct impacts on health due to heatwaves, dust or extreme weather; (2) changes to the agricultural and natural environment resulting in reduced food yields and undernutrition, changes to the water supply, and changes in climatic-sensitive infectious diseases; and (3) social and economic disruptions and hardships in climate-displaced groups, depression and despair common to marginalized, disadvantaged and failing farm communities, and the health consequences of tension and hostilities due to declines in food, water, and living space (McMichael 2012). Given the impoverished, malnourished condition and recently climate-displaced nature of most Europeans prior to 542 CE, it is not surprising that perhaps half of the population of the Byzantine Empire and Europe were killed between 542 and 565 CE by the Justinianic Plague, and perhaps as many as 100 million died (Russell 1968; Lamb 2002).

Arrival, spread, seasonality and epidemic velocity

The Justinianic Plague seems to have first reached Europe in the summer of 541 CE through Pelusium, Egypt (Findlay and Lundahl 2017; Cohn 2008), and all of the Eastern Mediterranean and North Africa were affected by 542 CE. By 544 CE and again in 549 CE it reached Ireland (Maddicott 1997; Cohn 2008). England was hit initially between 544–547 CE, and again in 634 CE (Retief and Cilliers 2006), and between 664–666 CE, 675–680 CE, 684–687 CE, throughout the 680s CE, and again in 715 CE, with great mortality, while Rome was hit in 667–668 CE (Maddicott 1997) (Figure 3.1).

Rome was hit again in 680 CE between July and September (Paul the Deacon, Little 2007). Ravenna, Grado and Istria were hit by plague in 593 CE (Paul the Deacon, Little 2007) suggesting that plague moved from the Turkish peninsula up into the Balkans and Italy the next year. A high-mortality plague outbreak hit southern France in 580 CE, again in 588 CE, in particular in Marseilles, and again in 590–1 CE (Allen 1979). A ship arriving from Spain was implicated in bringing the plague to Marseilles in 588 CE (Findlay and Lundahl 2017). Furthermore, the Middle East was hit by six further epidemics between 627 CE and 717 CE that spread as far west as the ports of Genoa, Italy, Marseilles, Gaul (France), and

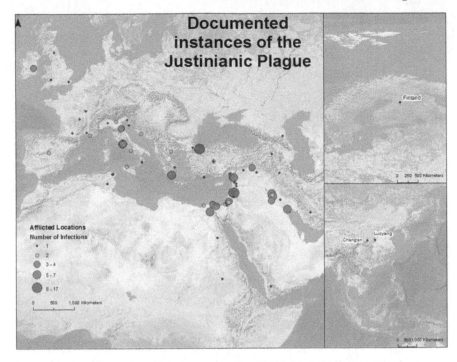

Figure 3.1 Documented instances of the Justinianic Plague (A. Davidson, 2011. Disputed Causality of the Justinian and Athenian Plagues: A Solution through Spatio-temporal Modeling. *Georgia Southern University B.S. Geography thesis.* Reprinted with permission from Andrew Davidson.)

Narbonne, Spain, before moving inland (Findlay and Lundahl 2017). Corippus notes that a terrible plague leading to societal breakdown arrived in North African by sea in 549 CE (Little 2007), and it returned again in 599–600 CE.

The average Justinianic epidemic velocity is ~1,112 kilometers per year, or 3 km/day. This is much higher than the 80 kilometers per year observed transmission velocity of modern bubonic plague among rodents in North America (Adjemian *et al.* 2007; Ari *et al.* 2010), while the medieval Black Death exhibited a transmission velocity of ~6 km per day (Benedictow 2004). However, in 1348, plague moved over a Euclidian, not network-derived, distance of 1,250 km (Bossak and Welford 2015), suggesting that Benedictow's (2004) number is likely accurate.

Although there is a strong correlation between afflicted locations and Roman trade routes (Figure 3.2), no real front of infection can be identified.

Furthermore, no seasonal pattern of infection and mortality can be detected for the Justinianic Plague, unlike the medieval Black Death that exhibited late summer and early fall infection peaks (Welford and Bossak 2009). The majority of the infection peaks (18) were in the summer, whereas only 14 occurred in winter. William McNeill (1976) declared that the epidemic was confined to the Mediterranean coast because the black rats that carried the plague were indigenous to India

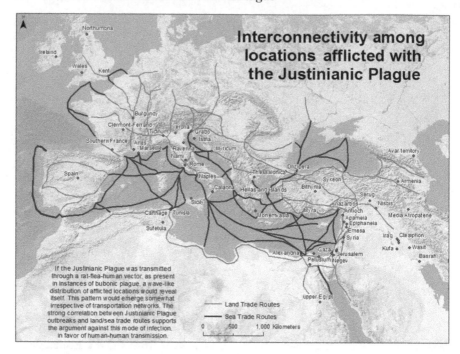

Figure 3.2 This map illustrates the interconnectivity among locations afflicted with the Justinianic Plague. (Reprinted from A. Davidson, 2011.)

and had not reached northern Europe during the time of Justinian, except by ship. These maps suggest that the epidemic was far more widespread than the coastal Mediterranean, thus invalidating this idea of seafaring black rats as central to the transmission of the Justinianic Plague. Interestingly, there is no evidence that rats were present during the time of the Athenian Plague; they were not endemic in classical Greece, and no terminology for rat exists in the Greek language (Retief and Cilliers 1998). If we accept these dates and locations of the Justinianic Plague, the epidemic velocities suggest that the Justinianic Plague was not bubonic but rather pneumonic plague, as suggested by recent genomic work (Cui *et al.* 2013; Harbeck *et al.* 2013; Zimbler *et al.* 2015; Wagner *et al.* 2014).

The Justinianic Plague possessed a remarkably high epidemic velocity, traversing approximately 556 kilometers between Constantinople and Sykeon along a Roman road between the spring and summer of 542 CE. If we allow six months for transmission, the epidemic velocity would be 1,112 km per year, independent of sea travel. Between fall 541 CE and summer 542 CE the Justinianic Plague moved from Alexandria to Pelusium to Askalon, a distance of 1,246 km land distance. These estimates are much higher than the 25 km a year traversed by modern bubonic plague. The vast area covered by the Justinianic Plague alone is indicative of human-human transmission, as the rat-flea-human vector is inefficient and naturally only reaches an average speed of 65 meters per day in the absence of human propagation (Welford and Bossak 2010). Moreover, locations afflicted with the Justinianic Plague were primarily large cities connected by Roman trade routes (see Figure 3.2). A human-human mode of transmission is consistent with commingling and travel along these trade routes, which can also explain the slight dominance of summer peaks of infection of this epidemic.

Mortality

Death tolls in Rome during the Antonine Plague reached 2,000 a day (Fears 2004), while in the Eastern Roman Empire during the Justinianic Plague, epidemic mortality was in excess of 25% (Cunha and Cunha 2008). These are comparable to death tolls observed in Europe during the medieval Black Death (Paine 2000). In Byzantium, modern Istanbul, the Justinianic Plague initially caused 5,000 fatalities each day, with mortality later rising to 10,000 each day or more (Procopius, Retief and Cilliers 2006; Maas 2005) and mortalities in the first outbreak in Constantinople of 56% (Allen 1979). Only one in three victims survived, with Justinian himself contributing to this statistic (Maas 2005). It is estimated that, by the end of the 6th century, fully a third of the Roman Empire had died. Yet northern and eastern tribes on the fringes of the Roman Empire, whose transportation systems were very limited and not tied to efficient land or sea routes of the Roman Empire, were hardly affected by either the Antonine or Justinianic plagues (Cunha and Cunha 2008). This spatial discontinuity in mortality suggests human-human transmission of both plagues was the norm rather than a non-human vector facilitating plague transmission. A non-human vector would hardly differentiate between human cultural groups across land masses.

Disease-induced mortality among Justinianic Plague victims also seems to exhibit a bias towards the poor (John of Ephesus; Retief and Cilliers 2006) with first 5,000, then 7,000, then 12,000 and ultimately up to 16,000 people dying on a single day, although the plague was only in its initial stages. This bias suggests social differentiation had a detrimental effect on health prior to the plague (Patlagean 1977). The heaviest toll of the 558 CE epidemic was among the young, while the Basra 706 CE outbreak was called the Plague of Maidens (Little 2007).

Etiologic identification

Like its predecessor the Athenian Plague, the onset of the Justinianic Plague (the first wave 541–544 CE and second wave 557–574 CE, Rosen 2010) was characterized by a sudden high fever that raged all night and day (Little 2007), swellings (bubos) of the groin, armpits, the neck, the thighs, and some patients developed a rash of black, pea-sized blisters (Procopius, Retief and Cilliers 2006), yet strangely it was reported that the victims were not hot to the touch. Coma, delirium, restlessness, insomnia, and visions were all reported (Procopius, Retief and Cilliers 2006). Domesticated and wild animals were also affected, and swollen tumors were found on dead cattle, dogs and even rats, among a host of others. It is noted that traditional burial rites were abandoned, a further indication of the huge scope of the plague. Areas escaping the effects of the plague included most of the Arabian Peninsula and central Asian steppes; they most likely lacked the population density necessary for its spread due to their largely nomadic populations (Maas 2005), or more simply, historical records have not been located that identify a plague in these regions.

It is essential to keep in mind that the texts of the Justinian Plague are rather vivid passages of historical writing rather than medical documents (Maas 2005). Taking this into account, we cannot place more than face value on statistics of mortality, and they are merely included for context rather than analysis. Writing in 2005, Michael Maas asserts that the historical diagnosis of plague was made not in a modern laboratory, but was essentially a judgment of the patient and his environment. Scholars of the Justinian Plague have traditionally agreed that it was an instance of bubonic plague (McNeill 1976), yet, recently, convincing arguments have been made by those studying the medieval Black Death against this classification (Maas 2005).

Given the epidemic velocity and mortality associated with the Justinianic Plague, some researchers have suggested that the Justinianic Plague was caused by the 1918 flu virus (Altschuler and Kariuki 2009) or by an emergent haemorrhagic virus (Duncan and Scott 2005). Nevertheless, recent anthropological laboratory work suggests that the Justinianic Plague was in fact a product of the *Yersinia pestis* bacterium (Drancourt *et al.* 2004; Harbeck *et al.* 2013; Wagner *et al.* 2014; Feldman *et al.* 2016). Skeletons from four adjoining mass graves from Sens, France, lack bone fractures, while indications of the sex and age of persons buried suggest they died due to an epidemic. These skeletons have been dated by radiocarbon to the 5th–6th century CE and have *Y. pestis* in their tooth pulp (Drancourt *et al.* 2004). Elsewhere, plague victims from two separate 6th century burials in southern Germany suggest the Justinianic strain of *Y. pestis* is Chinese in origin and has no modern descendants (Wagner *et al.* 2014; Feldman *et al.* 2016).

But the *Y. pestis* bacterium implicated is among a number of *Y. pestis* clones or biovars or strains (sub-species). Drancourt argues that the biovar *Antiqua* (Figure 3.3: O.PE3 strain) was responsible for the Justinian Plague, biovar *Medievalis* (Figure 3.3: Branch 2) from central Asia caused the plague from 1347–1801, while the modern pandemic is associated with the biovar *Orientalis* that has its geographic origin in east Asia (Drancourt *et al.* 2004). The plague strain implicated in the Justinianic Pandemic is believed to originate in north-east China (Figure 3.3 B; Cui *et al.* 2013). Although the Justinianic, medieval Black Death and modern bubonic plague lineages emerged in Central-East Asia, genetic coding indicates the Justinianic Plague represents a novel emergent event that does not give rise to later plague pandemics (Wagner *et al.* 2014). *Orientalis,* which is transmitted through the rat-flea-human route and hence is bubonic plague, is more distantly related (Drancourt *et al.* 2004). These quite different *Y. pestis* biovars could explain the different symptoms, epidemic velocities, and lethalities associated with *Y. pestis* endemics. Recent genomic work suggests, in fact, all known strains of *Y. pestis* have an origin in Central Asia (Harbeck *et al.* 2013).

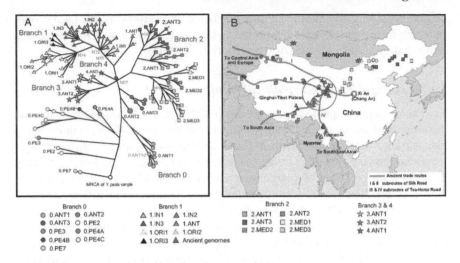

Figure 3.3 Global phylogeny and ancient trade routes for *Y. pestis*. (Reprinted with permission from PNAS from Cui et al. 2013.)

More recent work suggests five major genome branches for *Y. pestis*, that itself evolved from *Y. pseudotuberculosis* between 2,600–26,000 years ago, with the age of an *Angola* (Figure 3.3: O.PE3 strain) genome (possibly biovar *Antiqua*) overlapping with the Justinianic Pandemic (Cui *et al.* 2013). Furthermore, the acquisition of a single gene, pPCP1, encoding for protease Pla sufficient to cause pneumonic plague occurred very early in *Y. pestis* evolution among the ancestral strain of *Angola* and certainly before the Justinianic Plague (Zimbler *et al.* 2015). The I259 variant of Pla among *Y. pestis* causes mortality via pneumonia within the lungs, whereas several modern and ancient strains of *Y. pestis* have the I259T variant of Pla that allows a fully optimized invasive infection through the lymph nodes, and mortality is through sepis as in bubonic plague (Zimbler *et al.* 2015). This suggests the Justinianic Plague and the medieval Black Death are distinct from the third plague pandemic (Wagner *et al.* 2014). Both the Justinianic Plague and medieval Black Death were solely pneumonic in nature, while the later I259T substitution enhanced the invasive capacity of *Y. pestis,* leading to the third plague pandemic or modern bubonic plague.

The Justinian Plague is reported to have been much more lethal, with mortality figures approaching 20–30% or even as high as 40% (Little 2007), rather than the 1% or less exhibited by modern bubonic plague. Of the 66 SNPs (single nucleotide polymorphisms) identified among Justinianic Plague sequences, two *ail* and *yopJ* are strongly associated with virulence (Wagner *et al.* 2014). These figures are not concrete, however, so we suggest that arguments based on spatial patterns, transmission velocity, and seasonality are more relevant. With the help of modern steamships and railway trains, the bubonic plague in South Africa in 1899 only moved inland at 20 km per year (Maas 2005). This rate is much lower than the observed velocity of either the Justinian Plague or medieval Black Death: both of which have been associated with the two more closely related *Yersinia pestis* biovars: *Antiqua* and *Medievalis* (Drancourt *et al.* 2004).

Social/cultural/economic/political ramifications

The Dark Ages, precipitated by the break-up of the Eastern Roman Empire follow-ing the Justinianic Plague and Late Antique Little Ice Age (LALIA), witnessed the disintegration of towns, the failure of Greek and Roman-inspired public health sys-tems, and the cessation of practical, tested medicine practiced by Roman military physicians for hundreds of years previously. The 627–628 CE plague outbreak, known as the Plague of Sharawaygh, devastated Sasanian Iraq, killing the Sasa-nian ruler (Little 2007). In Northeast Asia, the Eastern Türkic Empire collapsed in 630 CE, although recent evidence points to a climatic cooling between 627–629 CE with associated frosts and snow killing horses and sheep (Fei *et al.* 2007). Nevertheless, the collapse of the Eastern Türkic Empire occurs suspiciously close to the 627–628 CE plague outbreak (Büntgen *et al.* 2016). On the other hand, Western Türks' expansion to the shore of the Black Sea encouraged Emperor Her-aclius of the Eastern Roman Empire to form an alliance with them: this alliance was instrumental in Heraclius's successful Persian gambit (Büntgen *et al.* 2016).

Contemporaries speak of many deserted villages, while archeological evidence points to the expansion of cemeteries and the cessation of new buildings across the Roman Empire (Retief and Cilliers 2006). Immediately following the initial 542 CE Justinianic Plague, agricultural activity collapsed, and again following the 556–557 CE plague, shortages of bread, wine and grain possibly led to extensive rioting in May 556 CE. However, attacks against minorities and/or those prac-ticing 'heresy' ceased, perhaps due to labor shortages (Retief and Cilliers 2006). Certainly between 543 and 717 CE, labor shortages immediately following plague outbreaks seem to precipitate financial crises across the Byzantine Empire, pos-sibly due to reductions in taxable income, while labor shortages in the Byzantine military certainly affected military campaigns (Hays 2005).

The loss of both manpower and revenue ended Justinian's hopes of reunifying both halves of the Roman Empire (Findlay and Lundahl 2017). From 543–717 CE, plague and LALIA reduced agricultural output, caused demographic collapse, ini-tiated massive food and consumer-good inflation, and ultimately paralyzed Euro-pean trade and commerce (Sabbatani *et al.* 2012a). But this is too simplistic; the impact of both plague and LALIA varied across space and time between 543–717 CE. In the 6th century, farmers in Anatolia moved toward a simpler nomadic pas-toralism (Haldon *et al.* 2014). In contrast, in the 7th century, farmers in the eastern Roman world introduced new crops, while the Eastern Roman Empire imposed a new tax and reorganized the military in the face of Arab-Islamic attacks (Haldon *et al.* 2014). It appears that in the face of Arab-Islamic attacks along the coasts of Balkans and Anatolia, the cultivation of vines and olives decreased, to be replaced by cereal production and pastoralism; however, an intensive agrarian society sur-vived adjacent to Constantinople (Haldon *et al.* 2014). Nevertheless, the plague and LALIA appeared to reduce Anatolian farming intensities, decrease both urban and rural populations, and precipitate the militarization of the Anatolian peninsula through the 6th, 7th and 8th centuries CE (Haldon *et al.* 2014). A re-intensification of agrarian society was not observed until the mid-10th century around Nazianzos in Cappadocia (Haldon *et al.* 2014).

The combination of the decline and fall of the Eastern Roman Empire that followed Justinian's death and the death of Maurice in 602, hydroclimatic fluctuations, droughts, and famines that peaked in the late 530s CE, and the arrival and associated mortality of the Justinianic Plague in 542 CE paralyzed European trade and commerce and initiated the Dark Ages (Sabbatani *et al.* 2012a). The Justinianic Plague may also have contributed to the emergence of the Islamic Empire (Dols 1977), as low population densities of nomadic tribes in the Saudi peninsula may have allowed them to avoid significant plague mortality between 627 and 716 CE and thereby expand into the void left by the decline and weakening of both the Eastern Roman and Persian Empires (Sabbatani *et al.* 2012b).

Migrations

Slavs moved into the Balkans, and Lombards into Italy, after plagues left demographic voids in these areas, while Arabs occupied Syria following high plague mortality (Paul the Deacon, Little 2007). Human remains found with healed wounds and skeletal lesions indicating a life of hard work in Ancient Nemea hint at the Slavic invasion of Greece in the 6th century CE (Garvie-Lok 2010). In contrast, Curta (2001) suggests that the formation of a distinct Slavic ethnic identity between 500 and 700 CE occurred immediately north of the Roman Danube frontier, with the earliest Slav settlement south of the Danube occurring after 700 CE. During Justinian's reign, the first Slav raiding parties were noted along the Danubian frontier adjacent to the eastern flanks of the Carpathians (Curta 2001). The weakening of the Roman Empire between 500 and 700 CE and the collapse of the Germanic Hun settlements further to the north and west in the Great Hungarian Plain possibly led to the physical expansion of the Slavic ethnic group or diffusion of their cultural and linguistic identities further west (Barford 2008). This occurred at a time when the Justinianic Plague led to declining tax revenues within the Eastern Roman Empire, which crippled state power and severely impeded the fitness of Roman's professional army. Mortality among Roman soldiers due to plague between 542–716 CE led to extensive recruitment of barbarians (e.g., Visigoths, Mongols) into the Roman army, but budget volatility frequently led to soldiers not being paid. This led to several army insurrections and the subsequent penetration of Slav raiding parties and Arab armies in the late 6th and early 7th centuries into the Eastern Roman Empire (Little 2007). These weaknesses and reductions in central authority led to an age of shifting religious affinities and an explosion of religious conflicts in the Middle East (Hays 2007). The expansion of the Slavic language can be traced to LALIA and the Justinianic Plague as eastern European peoples such as the Avars expanded into the void left by the ravages of drought, pandemic and Roman abandonment (Büntgen *et al.* 2016).

Religion

Nestorianism, a Syrian form of Christianity, spread east along the Silk Road complex and in 635 CE was introduced into China with the Tang Emperor Taizong of Tang (Li Shimin) commissioning the first Chinese Nestorian church. Nestorian Christianity declined after the Tang Dynasty, with followers restricted to western Chinese providences by the 1650s. From the 7th century, Arab Muslims traveled along the Silk Road to China. Together, the Nestorian Church's administrative network and Arab traders along the Silk Road may have facilitated the eruption of plague from Central Asia into Europe. Across Europe, paganism returned in the mid to late 6th century in an intermittent pattern. Bede, writing in the 700s CE, noted that people looked to idols or the devil for cures or used incantations and amulets to ward off the plague (Little 2007). In order to avoid plague, Arab armies in the Saudi peninsula, would return to Medina, establishing a pattern that would be repeated throughout the centuries (Little 2007).

Wages and employment

The increased cost of laundering clothes in 544 CE in Constantinople due to an acute labor shortage led to courtiers simplifying their dress (Little 2007). Justinian responded through an edict that returned prices and labor costs to pre-plague levels (Little 2007). In the late 6th and early 7th centuries in Spain and Italy, efforts to reduce the number of runaway slaves who were looking for salaried employment escalated, to no avail. By the end of the 7th century the cost of returning runaway slaves was such a burden to Visigoths that slavery ceased in Italy and Spain (Little 2007).

The high mortality of the Justinianic Plague triggered regional-wide increases in real wages in Europe and the Middle East (Pamuk and Shatzmiller 2014). The Justinianic Plague killed upwards of 9 million people in the Byzantine Empire or as much as 30% of its pre-plague population (Laiou and Morrison 2007). This precipitated an increase in the value of wage labor. In fact, the purchasing power of unskilled labor in Egypt and Iraq was significantly higher in the post-plague environment than before (Pamuk and Shatzmiller 2014). It appears that higher incomes coupled with lower agricultural goods prices and lower land rents in the post-plague environment facilitated higher per capita incomes, creating greater demand for high-end luxury goods, which contributed (among other factors such as political stability and security) to the Golden Age of Islam (Pamuk and Shatzmiller 2014). Evidence for this increased demand for luxury goods can be found in cookbooks and their increased preferences for imported spices from Central, East, and South-East Asia (Nabhan 2014; Pamuk and Shatzmiller 2014).

Environmental changes

In the rural fields in Liguria 565 CE, harvests were left abandoned, vineyards ignored, and pastoral places overgrown, while villages, towns and cities became refuges for wild beasts (Paul the Deacon, Little 2007). In central Turkey, pollen data recovered from lake sediments suggest significant reforestation between 670–950 CE (England *et al.* 2008).

Summary

The Justinianic Plague was a harbinger of the medieval Black Death that arrived in Europe in 1347. It ravaged Europe and the Middle East, killing upwards of 30% of their populations shortly after major climatic-reversals; it spawned region-wide increases in real wages, precipitated the break-up of the Eastern Roman Empire, destroyed Greek and Roman inspired public health systems, and terminated the use of practical medicine by Roman military physicians. In short, the Justinianic Plague plunged Europe into a long period of political instability, and economic and cultural isolation – the Dark Ages.

References

Adjemian, J.Z., Foley, P., Gage, K.L. and Foley, J.E., 2007. Initiation and spread of traveling waves of plague, Yersinia pestis, in the western United States. *The American Journal of Tropical Medicine and Hygiene*, *76*(2), pp.365–375.

Allen. P., 1979. The Justinian plague. *Byzantion*, *49*, 5–20.

Altschuler, E.L. and Kariuki, Y.M., 2009. Was the Justinian plague caused by the 1918 flu virus? *Medical Hypotheses*, *72*(2), p.234.

Ari, T.B., Gershunov, A., Tristan, R., Cazelles, B., Gage, K. and Stenseth, N.C., 2010. Interannual variability of human plague occurrence in the western United States explained by tropical and north Pacific ocean climate variability. *The American Journal of Tropical Medicine and Hygiene*, *83*(3), pp.624–632.

Bailhe, M.G.L., 1995. *A slice through time. Dendrochronology and precision*. London: B.T. Batsford Ltd.

Baker, A., Hellstrom, J.C., Kelly, B.F., Mariethoz, G. and Trouet, V., 2015. A composite annual-resolution stalagmite record of North Atlantic climate over the last three millennia. *Scientific Reports*, *5*.

Barford, P.M., 2008. Slavs beyond Justinian's frontiers. *Studia Slavica et Balcanica Petropolitana*, 2(4), pp. 21–26.

Benedictow, O.J., 2004. *The Black Death, 1346–1353: the complete history*. Suffolk, UK: Boydell & Brewer.

Blanchard, P., Castex, D., Coquerelle, M., Giuliani, R. and Ricciardi, M., 2007. A mass grave from the catacomb of Saints Peter and Marcellinus in Rome, second-third century AD. *Antiquity*, *81*(314), pp.989–998.

Boots, M. and Mealor, M., 2007. Local interactions select for lower pathogen infectivity. *Science*, *315*(5816), pp.1284–1286.

Boots, M. and Sasaki, A., 1999. "Small worlds" and the evolution of virulence: infection occurs locally and at a distance. *Proceedings of the Royal Society of London B: Biological Sciences*, *266*(1432), pp.1933–1938.

Bossak, B.H. and Welford, M.R., 2015. Spatio-temporal characteristics of the medieval Black Death. In *Spatial analysis in health geography*, eds. P. Kanaroglou, E. Demelle and A. Paez. Farnham, UK: Ashgate Press.

Bryce, J., Boschi-Pinto, C., Shibuya, K., Black, R.E. and WHO Child Health Epidemiology Reference Group, 2005. WHO estimates of the causes of death in children. *The Lancet*, *365*(9465), pp.1147–1152.

Büntgen, U., Myglan, V.S., Ljungqvist, F.C., McCormick, M., Di Cosmo, N., Sigl, M., Jungclaus, J., Wagner, S., Krusic, P.J., Esper, J. and Kaplan, J.O., 2016. Cooling and societal change during the late antique little ice age from 536 to around 660 AD. *Nature Geoscience*, *9*(3), pp.231–236.

Büntgen, U., Tegel, W., Nicolussi, K., McCormick, M., Frank, D., Trouet, V., Kaplan, J.O., Herzig, F., Heussner, K.U., Wanner, H. and Luterbacher, J., 2011. 2500 years of European climate variability and human susceptibility. *Science*, *331*(6017), pp.578–582.

Cheung, C., Schroeder, H. and Hedges, R.E., 2012. Diet, social differentiation and cultural change in Roman Britain: new isotopic evidence from Gloucestershire. *Archaeological and Anthropological Sciences*, *4*(1), pp.61–73.

Cohn, S.K., 2008. Epidemiology of the Black Death and successive waves of plague. *Medical History*, *52*(S27), pp.74–100.

Cui, Y., Yu,C., Yan, Y., Li, D., Li, Y., Jombart, T., Weinert, L.A., Wang, Z., Guo, Z., Xu, L., Zhang, Y., Zheng, H., Qin, N., Xiao, X., Wu, M., Wang, X., Zhou, D., Qi, Z., Du, Z., Wu, H., Yang, X., Cao, H., Wang, H., Wang, J., Yao, S., Rakin, A., Li, Y., Falush, D., Balloux, F., Achtman, M., Song, Y., Wang, J. and Yang, R., 2013. Historical variations in mutation rate in an epidemic pathogen, *Yersinia pestis*. *PNAS*, *110*, pp.577–582.

Cunha, C.B. and Cunha, B.A., 2008. Great plagues of the past and remaining questions. In *Palaeomicrobiology: past human infections* (pp.1–22), eds. D. Raoult and M. Drancourt, Berlin: Springer Verlag.

Curta, F., 2001. *The making of the slavs: history and archaeology of the Lower Danube Region, c. 500–700 (Volume 52)*. Cambridge: Cambridge University Press.

Davidson, A., 2011. Disputed causality of the Justinian and Athenian plagues: a solution through spatio-temporal modeling. B.S. Geography thesis, Georgia Southern University.

Diaz, H. and Trouet, V., 2014. Some perspectives on societal impacts of past climatic changes. *History Compass*, *12*(2), pp.160–177.

Dols, M.W., 1977. *The Black Death in the Middle East*. Guildford, Surrey: Princeton University Press.

Drancourt, M., Roux, V., La Vu Dang, L.T.H., Castex, D., Chenal-Francisque, V., Ogata, H., Fournier, P.E., Crubézy, E. and Raoult, D., 2004. Genotyping, Orientalis-like Yersinia pestis, and plague pandemics. *Emerging Infectious Diseases*, *10*(9), p.1585.

Dull, R.A., Southon, J.R. and Sheets, P., 2001. Volcanism, ecology and culture: a reassessment of the Volcán Ilopango TBJ eruption in the southern Maya realm. *Latin American Antiquity*, *12*, pp.25–44.

Duncan, C.J. and Scott, S., 2005. What caused the Black Death? *Postgraduate Medical Journal*, *81*(955), pp.315–320.

Duncan-Jones, R.P., 1996. The impact of the Antonine plague. *Journal of Roman Archaeology*, *9*, pp.108–136.

England, A., Eastwood, W.J., Roberts, C.N., Turner, R. and Haldon, J.F., 2008. Historical landscape change in Cappadocia (central Turkey): a palaeoecological investigation of annually laminated sediments from Nar lake. *The Holocene*, *18*(8), pp.1229–1245.

Fears, J.R., 2004. The plague under Marcus Aurelius and the decline and fall of the Roman Empire. *Infectious Disease Clinics of North America*, *18*(1), pp.65–77.

Fei, J., Zhou, J. and Hou, Y., 2007. Circa A.D. 626 volcanic eruption, climatic cooling, and the collapse of the Eastern Turkic Empire. *Climatic Change*, *81*(3–4), pp.469–475.

Feldman, M., Harbeck, M., Keller, M., Spyrou, M.A., Rott, A., Trautmann, B., Scholz, H.C., Päffgen, B., Peters, J., McCormick, M. and Bos, K., 2016. A high-coverage

Yersinia pestis genome from a sixth-century Justinianic plague victim. *Molecular Biology and Evolution, 33*(11), pp.2911–2923.

Findlay, R. and Lundahl, M., 2017. *The economics of the frontier*. London: Palgrave Macmillan.

Galvani, A.P., 2003. Epidemiology meets evolutionary ecology. *Trends in Ecology & Evolution, 18*(3), pp.132-139.

Garvie-Lok, S., 2010. A possible witness to the sixth century Slavic invasion of Greece from the stadium tunnel at Ancient Nemea. *International Journal of Historical Archaeology, 14*(2), pp.271–284.

Greenberg, J., 2003. Plagued in doubt. *Journal of Roman Archaeology, 16*, pp.413–425.

Haldon, J., Roberts, N., Izdebski, A., Fleitmann, D., McCormick, M., Cassis, M., Doonan, O., Eastwood, W., Elton, H., Ladstätter, S. and Manning, S., 2014. The climate and environment of Byzantine Anatolia: integrating science, history, and archaeology. *Journal of Interdisciplinary History, 45*(2), pp.113–161.

Harbeck, M., Seifert, L., Hänsch, S., Wagner, D.M., Birdsell, D., Parise, K.L., Wiechmann, I., Grupe, G., Thomas, A., Keim, P. and Zöller, L., 2013. Yersinia pestis DNA from skeletal remains from the 6th century AD reveals insights into Justinianic Plague. *PLoS Pathogens, 9*(5), p.e1003349.

Harper, K., 2016. The environmental fall of the Roman Empire. *Daedalus, 145*(2), pp.101–111.

Hays, J.N., 2005. *Epidemics and pandemics: their impacts on human history*. ABC-CLIO.

Hays, J.N., 2007. Historians and epidemics: simple questions, complex answers. In *Plague and the End of Antiquity: The Pandemic of 541–750* (p.33), ed. L.K. Little. Cambridge: Cambridge University Press.

Huntington, E., 1917. Climatic change and agricultural exhaustion as elements in the fall of Rome. *The Quarterly Journal of Economics, 31*(2), pp.173–208.

John of Ehesus, 1889. Joannis Episcopi Ephesi Syri Monophysitae Commentarii de Beatis Orientalibuset Historiae Ecclesiasticae Fragmenta. Latin translation by W.J. van Douwen and J.P.N. Land. *Transactions of the Royal Academy of the Netherlands, Literary Division 18*, Amsterdam: Müller.

Kühn, C.G. ed., 1821–1833. *Claudii Galeni Opera Omnia*. Leipzig: C. Cnobloch, rpt. Hildesheim: Georg Olms, 1964–5. (Greek, Latin trans.) *Editio Kuchniana Lipsiae*.

Laiou, A.E. and Morrisson, C., 2007. *The Byzantine economy*. Cambridge: Cambridge University Press.

Lamb, H.H., 2002. *Climate, history and the modern world*. London: Routledge.

Larsen, L.B., Vinther, B.M., Briffa, K.R., Melvin, T.M., Clausen, H.B., Jones, P.D., Siggaard-Andersen, M.L., Hammer, C.U., Eronen, M., Grudd, H., Gunnarson, B.E., Hantemirov, R.M., Naurzbaev, M.M. and Nicolussi, K., 2008. New ice core evidence for a volcanic cause of the AD 536 dust veil. *Geophysical Research Letters, 35*(4).

Little, L.K., 2007. Life and afterlife of the first plague pandemic. In *Plague and the end of antiquity: the pandemic of 541–750*, ed. L.K. Little. Cambridge: Cambridge University Press.

Littman, R.J. and Littman, M.L., 1973. Galen and the Antonine Plague. *The American Journal of Philology, 94*(3), pp.243–255.

Löwenborg, D., 2012. An iron age shock doctrine: did the AD 536–7 event trigger large-scale social changes in the Mälaren valley area? *Journal of Archaeology and Ancient History (JAAH)*, (4).

Maas, M. ed., 2005. *The Cambridge companion to the Age of Justinian*. Cambridge: Cambridge University Press.

McMichael, A.J., 2010. Paleoclimate and bubonic plague: a forewarning of future risk? *BMC Biology*, 8(1), p.108.

McMichael, A.J., 2012. Insights from past millennia into climatic impacts on human health and survival. *Proceedings of the National Academy of Sciences*, 109(13), pp.4730–4737.

McNeill, W.H., 1976. *Plagues and peoples*. New York: Anchor.

Maddicott, J.R., 1997. Plague in seventh-century England. *Past & Present*, 156, pp.7–54.

Marklein, K.E. and Fox, S.C., 2016. In morbo et in morto: transforming age and identity within the mortuary context of Oymaağaç Höyük, Northern Turkey. In *Theoretical approaches to analysis and interpretation of commingled human remains* (pp. 185–205). Cham, Switzerland: Springer International Publishing.

Nabhan, G.P., 2014. *Cumin, camels, and caravans: a spice odyssey* (Volume 45). University of California Press.

Paine, R.R., 2000. If a population crashes in prehistory, and there is no paleodemographer there to hear it, does it make a sound? *American Journal of Physical Anthropology*, 112(2), pp.181–190.

Pamuk, Ş. and Shatzmiller, M., 2014. Plagues, wages, and economic change in the Islamic Middle East, 700–1500. *The Journal of Economic History*, 74(1), pp.196–229.

Pang, K.D., Keston, R., Srivastava, S.K. and Chou, H.H., 1989. Climatic and hydrologic extremes in early Chinese history: possible causes and dates. *Eos*, 70, p.1095.

Patlagean, E., 1977. Pauvreté économique et pauvreté sociale à Byzance. 4e-7e siècles, EHESS. *Centre de Recherches Historiques. Civilisations et societes*, 48.

Retief, F.P. and Cilliers, L., 1998. The epidemic of Athens, 430–426 BC. *South African Medical Journal*, 88(1), pp.50–53.

Retief, F.P. and Cilliers, L., 2006. The epidemic of Justinian (AD 542): a prelude to the Middle Ages. *Acta Theologica*, 26(2), pp.115–127.

Roseboom, T., de Rooij, S. and Painter, R., 2006. The Dutch famine and its long-term consequences for adult health. *Early Human Development*, 82(8), pp.485–491.

Rosen, W., 2010. *Justinian's flea: plague, empire and the birth of Europe*. New York: Random House.

Russell, J.C., 1968. That earlier plague. *Demography*, 5(1), pp.174–184.

Sabbatani, S. and Fiorino, S., 2009. The Antonine Plague and the decline of the Roman Empire. *Le infezioni in medicina: rivista periodica di eziologia, epidemiologia, diagnostica, clinica e terapia delle patologie infettive*, 17(4), pp.261–275.

Sabbatani, S., Manfredi, R. and Fiorino, S., 2012a. The Justinian Plague (part one). *Le infezioni in medicina: rivista periodica di eziologia, epidemiologia, diagnostica, clinica e terapia delle patologie infettive*, 20(2), pp.125–139.

Sabbatani, S., Manfredi, R. and Fiorino, S., 2012b. The Justinian Plague (part two). Influence of the epidemic on the rise of the Islamic Empire. *Le infezioni in medicina: rivista periodica di eziologia, epidemiologia, diagnostica, clinica e terapia delle patologie infettive*, 20(3), pp.217–232.

Scheidel, W., 2002. A model of demographic and economic change in Roman Egypt after the Antonine plague. *Journal of Roman Archaeology*, 15, pp.97–114.

Sigl, M., Winstrup, M., McConnell, J.R., Welten, K.C., Plunkett, G., Ludlow, F., Büntgen, U., Caffee, M., Chellman, N., Dahl-Jensen, D. and Fischer, H., 2015. Timing and climate forcing of volcanic eruptions for the past 2,500 years. *Nature*, 523(7562), pp.543–549.

Simmonds, A., Márquez-Grant, N. and Loe, L., 2008. Life and death in a Roman city: excavation of a Roman cemetery with a mass grave at 120–122 London Road, Gloucester. Oxford Archaeological Unit Ltd.

Sohysiak, A., 2008. The plague pandemic and Slavic expansion in the 6th–8th centuries. *Archaeologiia Polona*, *44*, pp.339–364.

Strothers, R.B. and Rampino, M.R., 1983. Historic volcanism, European dry fogs, and Greenland acid precipitation, 1500 BC to AD 1500. *Science*, *222*, pp.411–414.

Wagner, D.M., Klunk, J., Harbeck, M., Devault, A., Waglechner, N., Sahl, J.W., Enk, J., Birdsell, D.N., Kuch, M., Lumibao, C., Poinar, D., Pearson, T., Fourment, M., Golding, B., Riehm, J.M., Earn, D.J., Dewitte, S., Rouillard, J.M., Grupe, G., Wiechmann, I., Bliska, J.B., Keim, P.S., Scholz, H.C., Holmes, E.C. and Poinar, H., 2014. Yersinia pestis and the plague of Justinian 541–543 AD: a genomic analysis. *The Lancet Infectious Diseases*, *14*(4), pp.319–326.

Watts, D.J. and Strogatz, S.H., 1998. Collective dynamics of "small-world" networks. *Nature*, *393*(6684), pp.440–442.

Welford, M.R. and Bossak, B.H., 2009. Validation of inverse seasonal peak mortality in medieval plagues, including the Black Death, in comparison to modern Yersinia pestis-variant diseases. *PloS One*, *4*(12), p.e8401.

Welford, M.R. and Bossak, B.H., 2010. Revisiting the medieval Black Death of 1347–1351: spatiotemporal dynamics suggestive of an alternate causation. *Geography Compass*, *4*(6), pp.561–575.

Wisburd, S., 1985. Excavating words: a geological tool. *Science News*, *127*(6), pp.91–94.

Zimbler, D.L., Schroeder, J.A., Eddy, J.L. and Lathem, W.W., 2015. Early emergence of Yersinia pestis as a severe respiratory pathogen. *Nature Communications*, *6*, p.7487.

4 The medieval Black Death arrives in Europe

The Second Plague Pandemic begins

In the middle of the 14th century, the Old World was rapidly transitioning to a small world environment, one that was increasingly highly connected and spatially integrated along the trade axis of the older overland Silk Road and the newly expanded maritime Silk Network of shipping lanes (Rezakhani 2010). Although these connections proved highly beneficial to the western world as they sought to catch up with the rich and technologically advanced China, the new embryonic small world environment that the Old World now found itself in, rather inconveniently, selected for greater disease transmissibility and greater disease virulence (Boots and Sasaki 1999; Boots and Mealor 2007).

For the first time, the east consistently traded and exchanged knowledge and technology and goods with the west. Both the overland Silk Road and maritime Silk Network acted as a conduit for trade, exchange of information and technology among China and Rome/Byzantium/Europe, and disease (Rezakhani 2010). The Silk Road/Network became the superhighways of their time. The commencement of trade along the overland Silk Road in the 2nd century BCE marked the beginning of the end of a large world, a world where people remained isolated and unknowing of each other, where local disease epidemics lived and died but were unable escape and infest other isolated communities. Thereafter, the Old World and its people would be forever linked by trade and disease transmission along the Silk Road/Network.

The overland Silk Road was a set of anastomosing paths and tracks that split and reconnected across the Central Asian steppes. Along the overland Silk Road, paced roughly a day's hike apart, were either villages, towns, or caravanserai, places where traders could find food, accommodation and shelter for themselves and their camels. The caravanserai, the villages, and the towns along the Silk Road all offered sites to trade goods. In fact, most goods moved westwards in a piecemeal fashion. Traders and their camel caravans would stop at Silk Road junctions, fortresses, oases, or even large caravanserai and conduct their business, trading their goods for other goods, gold, or money, and then return home. Few traders traveled the length of the Silk Road. As a result, goods, be they silk moving west, or gold or horses or wool moving east, took months, sometimes years, to complete the approximately 4,000 km trip. This stop/start trading slowed the propagation of disease along the overland Silk Road but it also allowed ample opportunities for opportunistic diseases to spread among and between humans and animals as they moved back and forth along the Silk Road. Unfortunately, the overland Silk Road also bisects the Asian Steppes, which is home to several mammals (i.e., voles, rats, gerbils, marmots) that can and do act as plague reservoirs.

Plague's poor timing

The timing of the primary wave of the medicval Black Death (mBD) was impeccable; it hit Europe between 1347–1353 at a time shortly after the end of the medieval warm period (750–1250) and the Great Famine of 1315–18 (Fraser 2011). Recent evidence from the 1944–45 Dutch famine suggests that famines compromise the general health and immune system of individuals for decades after the famine (Bryce *et al.* 2005; Roseboom *et al.* 2006). It also hit at a time when the gap between rich (monasteries and aristocracy) and poor (peasants) was growing, and population pressures on the land were leading to damaging land management practices (Fagan 2001; Fraser 2011). Between 1100–1300, the medieval age reached a zenith in terms of investment in cultural and artistic endeavors, population size, and land under cultivation, that was not reached again until the 1500s (Fraser 2011). Surpluses in food resulting from innovations in agricultural technology (the heavy plough) and forest clearance led to continental trade in food and timber products that ultimately deepened the pockets of the landholding class (the monasteries and nobles) and entrenched the feudal system (Fraser 2011). Efforts to squeeze more productivity from the serfs and land led to the ploughing of fragile, marginal upland soils and widespread soil erosion (Fraser 2011). Global climatic change initiated in 1258 with a volcanic eruption in Indonesia (Stothers 2000) and freshwater intrusion into the North Atlantic due to the melting of the Greenland ice cap during the medieval warm period (Mann *et al.* 2009; Shin *et al.* 2003) helped precipitate global cooling, which led to wild swings in seasons, temperatures and precipitation across the planet, culminating in the disastrous summers of 1315–18 (Fagan 2001; Lamb 2013). Clearly, then, European peasants (the vast majority of the population) were vulnerable, deeply impoverished, malnourished, and desperate (Fagan 2001; Fraser 2011; Herlihy 1997) and highly susceptible to the medieval Black Death. High mortality rates were evident in the 13th century and the higher-than-expected plague mortalities experienced across Europe between 1347–1353 suggest a pre-plague population in poor health (DeWitte 2015).

The medieval Black Death was also one of the first truly global pandemics, uniting east and west, a harbinger of things to come, such as the Spanish flu epidemic of 1918–19 and AIDS. Possibly originating in China in the 1330s (Bos *et al.* 2016; Schmid *et al.* 2015), it moved along caravan trading routes across Central Asia, once known to Marco Polo, before arriving with the Golden Horde at the gates of Caffa (Varlik 2008). To reiterate: *in a geographical sense, the medieval Black Death marked the beginning of a smaller, more highly connected, spatially integrated world where ideas, goods, and diseases could be rapidly transported to all corners of the globe*. At the same time, Europe was transitioning from a serf-dominated, rural society to an early interdependent, market-based capital-ist economy that connected towns and ports through transportation of goods and services. Yet as this economic transition was in full swing, medical knowledge remained mired in Hippocrates' and Galen's understanding of disease. Thus doc-tors of the time had a very limited ability to adopt any useful means to counter epidemics (Benedictow 2004, 2005). To repeat, a small world selects for pandem-ics and greater viral and bacterial virulence (Boots and Sasaki 1999; Bossak and Welford 2010). In a large world with extensive distances between human groups or clusters, low transmission rates and low virulence ensure that diseases neither infect nor kill all humans in a group or cluster, ensuring that the disease does not become extinct (Boots and Sasaki 1999). In a small world, one that is highly connected and where travel times between human groups or clusters are short, these local thresholds break down. As a result, highly virulent diseases can jump between clusters before local extinctions occur (Boots and Sasaki 1999). It would appear that in the 1340s, Europe and Central Asia reached some threshold of crit-ical connectivity among human clusters (Boots and Sasaki 1999). This critical connectivity ultimately killed as many as 50 million people or 60% of Europe's population (Benedictow 2005) and millions in Central Asia, when the plague underwent a cross-species transmission from marmots to humans (Bos *et al.* 2016; Schmid *et al.* 2015; Stenseth *et al.* 2006; Wilschut *et al.* 2015) in the central steppes of Asia somewhere in the late 1330s or early 1340s and then spread along the Silk Route.

What is the origin of plague and is it the medieval Black Death?

The Asian steppes that stretch from Kazakhstan to China are home to several mam-malian plague reservoirs. These include the great gerbil (*Rhombomys opimus*), a rodent that periodically suffers plague epidemics (Stenseth *et al.* 2006; Wilschut *et al.* 2015); the Yunnan red-backed vole (*Eothenomys miletus*; the buff-breasted rat (*Rattus flavipectus*); and two marmot species (*Marmota himalayana, Marmota baibacina*) (Eppinger *et al.* 2009). For instance, in Kazakhstan, eruptions of plague among gerbils are driven by regional climate phenomena. In cool, dry springs and summers on the steppes of Kazakhstan, gerbils struggle to survive, as grass forage is limited. As a result, they tend to live in disconnected colonies that do not support plague epidemics. In contrast, in warm, wet springs and summers, gerbils thrive, populations grow, and colonies once isolated expand and interact with one another, providing a perfect environment for plague to erupt and be transmitted via a flea vector between burrows and colonies across the steppes (Stenseth *et al.* 2006; Wil-schut *et al.* 2015). The horizontal transfer of plague among burrows, known as a plague eruption, occurs when more than 50% of gerbils in any one burrow or col-ony are plague-positive (Stenseth *et al.* 2006). Although this dynamic has not been identified in the other plague reservoirs in Asia, it is not hard to visualize a similar plague eruption among voles, rats, or marmots.

It is generally agreed that to seed permanent cross-species transmission of any disease from mammals to humans a series of contingent conditions need to be met until a self-sustaining population of infected humans occurs. Simply put, once plague clusters within mammalian populations became pandemic eruptions, contact with humans will escalate. Cross-species transmission probably occurred among those hunting these mammals for food, or when these mammals under food stress invaded human-cultivated fields and homes during subsequent falls and winters when forage became scarce. At some point in the early 1330s, someone (or several people) contracted the plague. I suspect several or nearly all of these cross-species transmissions did not result in anyone other than the initially infected individual or their immediate family succumbing to the plague. But somewhere out on the Asian steppes an individual, a small family group, or an itinerant trader transmitted plague to other humans. Again, this leap could have remained local, with the plague killing most of a village or region, but at that time in the 1330–40s, the Silk Road/Network and Golden Horde military expansion provided the means for plague to infect, spread, and kill indiscriminately across Asia and move on toward Europe. Each step is contingent upon the prior step, but once plague became self-sustaining among the people and traders, sometime between 1282–1343 within the Asian steppes, it took plague just 10–14 years to propagate along the Silk Road/Network to Europe (Bos *et al.* 2011; Schmid *et al.* 2015).

Which plague strain seeded each plague pandemic?

Plague strains isolated in China seem to identify the source of the three plague pandemics (Morelli *et al.* 2010). Phylogenetic diversity among a global collection of 286 *Y. pestis* isolates suggests *Y. pestis* evolved in China and seeded multiple epidemics via the Silk Road or Chinese marine voyages (Morelli *et al.* 2010). The plague strain (E1979001) isolated from voles is associated with the biovar *antiqua* (the Justinianic plague strain). The rat plague strain (F1991016) is from the biovar *orientalis* (the modern plague strain), and marmots exhibit two strains (K1973002 and B42003004) associated with both *antiqua* and *medievalis* (the medieval Black Death) biovars (Drancourt *et al.* 2004; Eppinger *et al.* 2009; Zhou *et al.* 2004). The most ancient *Yersinia pestis* strain lineage in China is B42003004 and is genetically closest to the original species, *Yersinia pseudotuberculosis* (Eppinger *et al.* 2009). More recent work suggests that all human *Yersinia pestis* strains evolved from a common ancestral strain (B42003004, derived from marmots) between 1282–1343 CE in East Asia (see Figure 3.3; Cui *et al.* 2013). In fact, multiple strains representing four branch lineages (Branches 1, 2, 3, and 4) of *Yersinia pestis* diversified in central Eurasia in what has been called the 'Big Bang' several decades before arriving in Europe (Figure 3.3; Bos *et al.* 2011; Cui *et al.* 2013; Green and Schmid 2016).

These four strains persisted and circulated in London and Europe in the 14th century (Bos *et al.* 2016; Cui *et al.* 2013; Haensch *et al.* 2010). This suggests that four strains of *Yersinia pestis* derived from marmots moved down the Silk Roads/Network or with the Golden Horde to Caffa by 1346 and then into Europe. Of these four lineages, one erupted in Provence between 1720–1722 (Bos *et al.*

2016). It is possible that the multiplicity of plague strains circulating in Europe between 1346–1353 increased plague mortality because Europeans were unable to attain herd-acquired plague immunity as each strain triggered slightly different antibody responses. However, all human *Yersinia pestis* strains exhibit few changes to the genes that code for virulence. This suggests that other factors such as climate, vector dynamics, human socio-economic conditions and possibly co-infection with other diseases might contribute to plague virulence (Bos *et al.* 2011). The 'Big Bang' in plague diversification was possible because although plague genes are monomorphic, single nucleotide polymorephisms (SNPs) (mutations in DNA code) accrue at the rate of 1 SNP every 8 years among plague genes (Green and Schmid 2016). Analysis of medieval plague DNA taken from human remains found in London suggests that the 1361–1363 outbreak differed (London sample 6330) from the 1346–1353 initial outbreak of medieval Black Death by as many as 3 SNPs (Spyrou *et al.* 2016). Elsewhere, a sample from Bolgar City, Russia, dated to 1362–1400 has two new SNPs but is missing one of sample 6330 SNPs (Green and Schmid 2016). This supports the notion that the plague of the primary wave of the medieval Black Death (1348–1353) was genetically different from the next outbreak of plague between 1361 and 1363, and this might explain their differences in age-related mortality (Green and Schmid 2016). All told, 97 SNPs have accrued on the plague genome since 1348 (Bos *et al.* 2011).

Which plague was it – bubonic or pneumonic plague?

There is neither archaeological nor narrative evidence that supports bubonic plague; in fact, there is no evidence – *description, narrative or physical evidence* – during the entire medieval Black Death in Europe for one or more rodent epizootic events (Cohn 2008). (An epizootic disease is a disease that is temporarily prevalent and widespread in an animal population.) Neither the diarist Samuel Pepys in England nor Absalon Pederssøn in Norway describe any rat die-off before an mBD/plague pandemic (Hufthammer and Walløe 2013). Rats, especially black rats whose fleas can carry *Y. pestis*, were rare to non-existent throughout Europe between 1347–1859 (Amori and Cristaldi 1999; Davis 1986; Rielly 2010; Schmid *et al.* 2015). The urban black rat (*Rattus rattus*) population densities in the UK between 1347–1351, identified from rat bones discovered in medieval rubbish tips, were not sufficient to sustain a rat-flea-human epidemic (Davis 1986; Rielly 2010; Sloane 2013). Black rats, first documented in the UK shortly after 43 CE, were present in the UK in the early medieval period, but suffered population crashes in both the 7th and 10th–12th centuries (Sloane 2013). Comprehensive archaeological analysis could find no evidence for rats in London between 1347–1351 (Sloane 2013; Rielly 2010). Furthermore, rural-dwelling brown rats (*Rattus norvegicus*) only reached western Europe from Russia in the early 18th century (Twigg 1984; Amori and Cristaldi 1999; Suckow *et al.* 2005), while today black rats (*Rattus rattus*) only persist in the UK due to recurrent introductions in and around ports of entry (Davis 1986) and large urban areas (Duncan and Scott 2005). In fact, there is no evidence for rats in Iceland until the 1800s, but two plague epidemics killed thousands in the 15th century (Karlsson 1996; Cohn 2002; Duncan and Scott 2005). An extensive examination of 120 collections, dating from 900 CE to the present, of bone material from Norway found rat bones in just 19 collections, and most of these were from the coastal towns of Oslo, Tonsberg, Stavanger, Bergen and Trondheim (Hufthammer and Walløe 2013). Moreover, the rat-flea-human transmission of plague is slow and inefficient. Vector-borne bubonic plagues in the United States have been averaging velocities of just 25 km/year since 1900 and 13–19 km/year between 1866 and 1994 in China (Benedict 1996). In contrast, during the primary wave of the medieval Black Death (mBD) in 1347–1353, the disease propagated between 0.9 and 6 km/day (Benedictow 2004; Christakos *et al.* 2005). We can conclude, then, that from the low density of black rats, the lack of rural brown rats, the rapid transmission velocities, and the high human mortality observed during the medieval Black Death from 1347–1859, a rat-flea-human transmission seems highly unlikely, if not impossible.

Pneumonic plague, a form of *Yersinia pestis*, is spread without the need of a flea as a vector and has been proposed as an alternate cause for the epidemics (Duncan *et al.* 2005; Haensch *et al.* 2010). Although modern symptoms of pneumonic plague do not match all the symptoms seen in historical documents, recent work suggests that the three *biovars* do vary in their ability to cause invasive infections. Nonetheless, the Novgorod chronicle of 1352 clearly stated that "a man would spit blood..." (Alexander 2002). *Biovar Antiqua* associated with the Justinianic Plague carries on its pPCP1 plasmid the code for a protease Pla that is conserved among all modern strains of *Y. pestis*. This single mutation separates this organism from *Y. pseudotuberculosis* (Zimbler *et al.* 2015). This protease Pla allows *Y. pestis* to induce severe pneumonia but not an invasive infection like bubonic plague. Recent changes to Pla among modern *Y. pestis* strains have optimized protease activity that allows fully invasive infections in humans (Zimbler *et al.* 2015), thus *Y. pestis* has evolved from a pneumonic infection to a bubonic infection. So, sometime and somewhere between 1282–1343 CE when the ancestral strain of *Y. pestis* evolved (Bos *et al.* 2011) and today, *Y. pestis* changed its stripes. Plague transitioned from a human-human transmissible disease to a flea-rat-borne disease.

In fact, the strain of plague that is responsible for the third (modern) plague pandemic appears to be quite distinct from the second pandemic plague, suggesting it survived and radiated in East Asia as a bubonic plague (Bos *et al.* 2016; Morelli *et al.* 2010). Although at times pneumonic infections erupt, these quickly exhaust themselves as they kill very quickly with high fatality (Ratsitorahina *et al.* 2000). So although transmitted among Central Asian rodent species such as great gerbils or marmots either as pneumonic or bubonic plague, once it crossed species in the 1340s the plague was pneumonic among humans and thus primed to devastate Europe and China as the medieval Black Death (mBD).

Arrival in Europe

Medieval Black Death was first recorded in Europe in the port city of Messina, on the Mediterranean island of Sicily, in October 1347, when a Genoese fleet full of sick and dying sailors arrived in port (Gottfried 1993). The traditional story of plague's arrival in Europe begins with the siege of Kaffa (now Feodosia, Ukraine), where Genoese traders and sailors were infected with the mBD by the Mongol Golden Horde and its leader Janibeg in 1346–47 (Wheelis 2002). After the Mongol army attacking Kaffa was decimated by the mBD, decaying corpses were reportedly catapulted over the city walls. Gabriele de' Mussi's account of this activity in 1347 hints at a causal association between the vaulted corpses and subsequent infection of the Genoese traders within the city of Kaffa. Three months after the terrified Genoese fled Kaffa by ship for home in Italy, the disease reached mainland Europe.

In contrast to this traditional narrative that Benedictow (2004) believes is a myth, Ibn al-Wardi, an Arab historian, reports that Muslim merchants witnessed plague in what is modern-day Uzbekistan in October and November of 1346 and observed its transmission to Crimea and Byzantine Constantinople (Ibn al-Wardi d. 1349; Varlik 2015). Although both narratives suggest Crimea as a source of the European plague, the latter narrative's mention of Constantinople offers an intriguing and highly likely location for plague's transmission from Asia to Europe. Another contemporary chronicler, Giovanni Villani, suggests Genoese merchants contaminated Constantinople and southern and central Greece by early July 1347 before moving onto Messina (Benedictow 2004; Schevill 1961; Varlik 2008). Equally conceivable is that the Genoese fled Kaffa without being infected with plague, only to subsequently contract plague when they stopped off in Constantinople on their way to Sicily and Ragusa. Irrespective of where plague originated, once it got to Sicily it then took less than a year for the mBD to reach the United Kingdom (Benedictow 2004).

The generally accepted path that mBD took after leaving Kaffa in late 1347, aboard Italian merchant ships, was that it reached Messina in October 1347 (Benedictow 2004; Snell 2017). Other Genoese fleeing Kaffa sailed up the Adriatic, first infecting Split on December 25, 1347 then Ragusa (modern-day Dubrovnik) by January 13, 1347, and then Venice by January 25, 1348, while elsewhere mBD reached Alexandra and Cairo by November 1347 (Benedictow 2004). According to Villani, shortly after contaminating Messini, inhabitants fled inland, contaminating Catania and, by December 1347, Marseille (France), Sardinia, Corsica and Elba Island (Benedictow 2004). In the Near East, mBD advanced just as quickly, contaminating Constantinople by the summer of 1347, and reached Tivrik in Armenia by September 10, 1348 (Varlik 2008). By the end of December 1347, mBD had crossed to Reggio di Calabria, and by January 1347, Pisa, while in France mBD had reached Aix en Provence some 25 km north of Marseille (Benedictow 2004). Thereafter, the disease spread inland reaching Avignon, (possibly by boat) Barcelona, Valencia, and Seville in Spain. In spring 1348, trading ships leaving Bordeaux contaminated Santiago de Compostela, La Coruna, Rouen, Weymouth and Melcombe (Benedictow 2004). *Grayfriars' Chronicle* states that mBD arrived in Melcombe on June 25, 1348 from one of two ships arriving from the continent via Bristol. Once in the Mediterranean and coastal countries of western Europe, mBD followed a hierarchical pattern of transmission following major European trading routes; it first spread to seaports, cities, and commercial hubs, then moved along navigable rivers, then radiated to regional towns, then reached smaller local market towns, and finally traveled into the countryside (Benedictow 2004; Harrison 2000; Mackay 1997). By December 1348, much of western Europe was in the grip of the disease. In July 1349, mBD swung slowly eastward and jumped to Bergen, Norway. The plague struck Russia in 1351–52 at Pskov and Novgorod before disappearing, only to return in the second wave of mBD that moved up along the Volga in 1364–65 (Alexander 2002).

The primary wave (the first epidemic) of the mBD was by far the most devastating, destroying the burgeoning economy of the late Middle Ages and killing more than 30% of the European population in a very short period of time; the true number of case fatalities and the actual mortality rate from this epidemic is still unknown, and estimates are derived from the postulations of individual historians (Horrox 1994; Benedictow 2004; Kelly 2005). The writings of Gabriele de' Mussi, a contemporary Italian chronicler of the disease and its societal effects, note that people throughout Europe, and indeed people of numerous cultures, viewed the mBD and its effect on people as a sign of the end of the world. De' Mussi's observations of popular thought at the time were correct in that the mBD represented a dramatic transition between the end of the Middle Ages and the dawn of the Renaissance (DesOrmeaux 2007; Herlihy 1997). In an interesting parallel, the Justinian Plague pandemic of 541–750 marked the end of Antiquity and the beginning of the Middle Ages (Little 2006).

Speed of transmission

The medieval Black Death was transmitted north-west, north, and east in Europe at an average rate of 0.9 to 6 km a day (Noble 1974; Benedictow 2004; Christakos *et al.* 2005; Bossak and Welford 2015). The largest and most detailed compilation of hundreds of individual mBD accounts illustrates that mBD expanded west and north-west in 1347 and 1348, and north-east in both 1349 and 1350, while the average disease transmission distance during 1348 was 1,250 km (Bossak and Welford 2015). Several factors could explain the initial rapid north/north-west expansion from Messina. First, a high-density road network based mostly on Roman roads served northern Italy, the Rhone valley, and the northern slope of the Pyrennes at this time. Second, this road network connected many large urban centers and a substantial number of ports. For instance, Santiago de Compostela became a pan-European place of pilgrimage beginning in the 10th century (Fletcher 1984), and routes across northern Spain to Paris during the 1300s were well served with roads, paths, and sites for lodging. The many large rivers that bisect the European Plain, and the high plateau of Spain, and the fact that Britain is an island, did not present much of an obstacle to the transmission of mBD. To reiterate, medieval Black Death or plague moved very rapidly in a wave-like propagation across Europe, initially toward the west, then arcing north and finally north-east and east, with the result that within three years from entry into Italy it had reached Gotland, Sweden (Bossak and Welford 2015).

In contrast, in the mid-1300s, central and eastern Europe was served with a lower-density road network and had limited access to ports. Only the Rhine valley was a major corridor of trade, and it was largely north-south, not west-east. In contrast, since its introduction in the US in 1900, bubonic plague has spread across the western US at a maximum of 80 km/year, while natural obstacles such as rivers and mountains have hindered plague expansion (Adjemian *et al.* 2007).

The mBD was a truly lethal global pandemic, spreading widely over most of the known world of the time, persistent in multiple natural environments and climatic spheres, and transmissible along the chain of human contact throughout all of the existing trade networks of the time. Only the lack of existing trade routes across oceans or within Eastern Europe and inhospitable environments limited its spatial range of affliction, although it penetrated alpine valleys with ease.

The mBD tended to move slowly across land but jump ahead via contaminated shipping vessels. Benedictow (2004) argues that medieval shipping could achieve and sustain daily sailing distances of 40 km/day, while land trade (i.e., riding horses or walking) led to sustained speeds of 0.5–2 km/day. For instance, mBD moved 160 km from Pisa to Pistoia, Italy at a pace of 1.78 km/day while advancing between Lyon and Givy in France at 1.2 km/day (Benedictow 2004). Along the Rhine, daily velocities of 2.2 km/day were achieved as mBD advanced from Basle to Mainz to Cologne in 7½ months (Benedictow 2004). Elsewhere, away from the main roads, the pace of mBD slowed; for instance, between Geneva to Lausanne it dropped to 0.66 km/day and between Geneva to Sion to 0.75 km/day (Benedictow 2004). Interestingly, high population densities do not seem to correlate with high mBD transmission velocities. Rather, mBD velocities appear to correlate with the connectivity of a specific location or region or country. For instance, the UK, Norway, Sweden and Denmark were conquered in one plague season, and mBD did not persist threatening populations for several years, as was the case in Italy and France, which reported plague in 1352 (Benedictow 2004). The strong internal trading networks and highly connected local and regional markets of the UK and Scandinavia appear to have facilitated rapid transmission, while more socially and politically fragmented regions with high rural population densities, such as Italy and France, appear to have slowed transmission velocities.

Transmission speeds exhibited seasonal patterns; even in southern France where winters only last 1–2 months, mBD was stopped or slowed down significantly (Benedictow 2004; Biraben 1975; Carpentier 1962; Dubois 1988). Elsewhere in the UK, Germany, and Scandinavia, where mBD arrived in the late autumn, expansion to the rest of the region only occurred in the following spring and summer (Benedictow 2004; Olea and Christakos 2005; Welford and Bossak 2009). It would appear that longer winters in northern Europe effectively stopped the transmission of mBD, while in southern Europe winters typically only slowed mBD contamination.

Spread: where, where not and attempts at control

Several factors could explain the initial rapid north/north-west expansion from Messina. First, a high-density road network based mostly on Roman roads served northern Italy, the Rhone valley, and the northern slope of the Pyrenees at this time. Second, this road network connected many large urban religious and commercial centers and a large number of ports. The old Roman roads and religious pilgrimage road networks focused on most of the great medieval commercial, trading, and religious cities and ports of Europe, and these cities, such as Santiago de Compostela, acted as 'bridgeheads' for mBD to spread and infect the interior of Europe (Benedictow 2004, 2005). Moreover, once in the UK, mBD swept across England in less than a year, in part because England had the densest Roman road network, second only to Italy. Third, western European trade was dominated by local and international shipping. Living as we do in the early 21st century – when you can order a personalized, engraved iPhone from Apple and have it shipped to you in 72 hours from Shanghai, China – we tend to forget that in the 1300s, Europe, especially north-west Europe (the British Isles and Flanders) and northern Italy, was transforming from a serf-dominated, rural inward-looking society to an early interdependent, industrial, market-based capitalist economy stretching from China to Europe along the Silk Road and around Europe, via busy shipping lanes. These shipping lanes linked, among many others, the Venetians and Genoese with Constantinople, Kaffa, and the end of the Silk Road; the Venetians and Genoese with London and Brugge; and London and Brugge with the German Hanseatic League (Benedictow 2004, 2005). In fact, England was initially infected through Melcombe, Weymouth, and Bristol – ports in south-west England. At the time, Italy was still suffering from the mBD when the mBD reached Weymouth in the late autumn of 1348 (Benedictow 2004, 2005). Even within the UK, spatial variations in mBD transmission were commonplace.

> If one tries to chart its [mBD] movements in detail it becomes obvious, as indeed one would expect, that it did not advance evenly in a broad swath across the country. For instance, in spite of the plague's early arrival in Bristol and Dorset, in England, rural Devon immediately to the west of both Dorset and Bristol seems not to have been affected until the following year.
>
> (Horrox 1994)

The density of rural markets and the connectivity to transmission pathways also affected plague transmission and mortality and illustrates the notion of medieval centrality. The county of Suffolk in England, for example, was severely affected by mBD in 1348 with mortalities exceeding 50% (Bailey 2010; Lock 1992; Poos 1991). This is in all likelihood because Suffolk had the highest density of rural markets in England at ten weekly market sites per 100 mi^2. This is nearly three times the national UK average of 3.7 in 1348 (Bailey 2010). At the time, Suffolk was also one of the richest agricultural areas in England. Conversely, in Spain, six isolated, poor rural Navarra communities experienced less than

5% plague mortality (Berthe 1984). Rich, populous, highly connected medieval market towns, capitals, and ports suffered higher mortalities and longer epidemics (Olea and Christakos 2005) than poorer, smaller, isolated rural communities across Europe. This suggestion is supported through a GIS comparative analysis of medieval transportation (excluding rivers) and plague outbreaks that found that 222 out of the 267 primary-wave medieval Black Death (1346–1353) localities (or 83.2%) were within 8 km of a transportation network (including port cities), while 92.9% were located within 24 km (Bossak and Welford 2015).

Simply put, a large population has more susceptible individuals who will sustain a longer epidemic, and a large, well-connected market town or port or capital will suffer more instances of infected individuals arriving in that town, port, or capital to conduct trade, political business, or religious activity than a smaller less diversified town. Moreover, the most connected urban locations were the first infected, while the least connected locations in northern and eastern Europe were the last to suffer the scourge of the medieval Black Death (Bossak and Welford 2015). Either the exclusion or dearth of medieval traders and pilgrims could account for the absence of either medieval Black Death or for those regions being the last to contract the plague. Certainly, those locations off the beaten trail were more likely to be on the lookout for ill travelers or restrict entry to foreigners. While the North/South divide created by the Rhine, the legacy of Roman roads populating western Europe, and the distance in central and eastern Europe from coastal ports could have protected eastern Europe from early inundation of the region by the plague (Bossak and Welford 2015).

It would appear, then, that geographic proximity to trade routes seems to have controlled the severity of plague mortality during the primary mBD epidemic wave (1347–1353). In fact, mBD mortality appears to exist in a cline from a peak along the most heavily traveled trade routes, many of which descend from original roads created during the Roman Empire, to a nadir along the tertiary and quaternary trade routes of the time (Bossak and Welford 2015). High mBD mortalities were observed in ports, for instance Venice (60–70% mortality; De Mussis, Gottfried 1993), Great Yarmouth, UK (>70%; Britnell 1994), and Montpellier (39–90% mortality; Wood and DeWitte 2003). Religious pilgrimage sites such as Santiago de Compostela and routes to these centers were also hard hit, such as Ribera near Estella that suffered mortalities of 67% (Zabalo Zabalegui 1968; Benedictow 2004). Routes crossing the northern Pyrenees, at two points tied to the old Roman road network, are associated with high mortalities. However, many villages in Navarre, Spain, just off these routes, were barely touched by mBD (Berthe 1984). These include Areso, Arriba, Arraiz, El Busto, Narbatte, the district of Olaibar, and Suarbe, which suffered no mBD deaths (Berthe 1984). However, Baigorri suffered 25% mortality and Pamplona only 16% mBD mortality, yet both were on these pilgrimage routes (Berthe 1984). On the Balearic Islands, which initiated a quarantine, mortalities were also very low, with the town of Inca suffering only 20% mBD mortality (Benedictow 2004; Christakos *et al.* 2005). Although isolated montane communities in Navarre, Spain avoided mBD, in the Northern French Alps mortalities exceeded the rest of Europe (~45%) (Gelting 1991).

Quarantine begins

Efforts to control the spread of plague and mitigate its effects marked the beginning of public health activities (Slack 1988). There are accounts of cities and towns such as Milan closing themselves off from the rest of the world in 1348, either not letting anyone in or out, or quarantining the infected individuals in their dwellings (Benedictow 2004; Hecker 1838; Slack 1988). In Milan, when cases of the plague were first discovered, all of the occupants of the three houses concerned, dead or alive, sick or well, were walled up inside and left to perish (Hecker 1838). In 1348, Venice appointed Nicolaus Venerio, Marinus Querino, and Paulus Belegno as overseers of public health. These officers were authorized to spend public money for the purpose of isolating infected ships, goods, and persons at an island in the lagoon (Eager 1903). In Pistoia, Italy in 1348, the city council set up 23 statutes aimed at controlling the mBD. These include Statutes I and II (taken directly from Chiappelli 1887):

I. No citizen of Pistoia or dweller in the district or the county of Pistoia ... shall in any way dare or presume to go to Pisa or Lucca or to the county or district of either. And that no one can or ought to come from either of them or their districts ... to the said city of Pistoia or its district or county on penalty of £50.

II. No person whether citizen, inhabitant of the district or county of the city of Pistoia or foreigner shall dare or presume in any way to bring ... to the city of Pistoia, its district or county, any used cloth, either linen or woolen, for use as clothing for men or women or for bedclothes on penalty of £200.

These two statutes clearly set out the beginnings of a quarantine regime.

In Salé, Granada, Ibn Abu Madyan walled up himself, his household, and a plentiful supply of food and drink and refused to leave the house until the plague had passed (Ibn al Khatib 1369 in al-Khanji 1978). He was entirely successful, although it is not clear this occurred in the primary wave of mBD. In April 1348, Pedro IV instructed the Government of Majorca to take steps to prevent the further propagation of mBD (Arandez 1980; Tuchman 1978) which seems to have been successful as Majorca only suffered a mortality of 16%. In fact, the word 'quarantine' descends from the Italian words for 'forty days', a period of time long enough to determine whether a sequestered individual was infected or not during the mBD (Scott and Duncan 2004). The cities and towns that followed precautions such as quarantining suspected plague victims demonstrated lower mBD mortality (if any was recorded at all) (Scott and Duncan 2004).

The emergence of public health regulations in Venice and Pistoia in 1348 represent the beginnings of the growth and expansion of local and state powers to restrict individual liberties in times of crisis (Slack 1988). By the 1720 plague outbreak in Marseille, entire cities were cordoned off (*cordons sanitaires*) and access to them controlled by military force (Slack 1988). By the 1600s, many

Mediterranean ports were cooperating to monitor suspect ships coming from the Middle East (Biraben 1975; Slack 1988).

Mortality, birth rate and seasonality

Overall mortalities for the UK and mainland Europe vary between 30–35%. However, Norway seems to have been particularly hard hit, with general population mortalities between 40–50% (Oeding 1990). Elsewhere, only in rural Cambridgeshire, Suffolk and Essex in 1348 in England did mBD mortalities consistently exceed 50% (Aberth 1995; Bailey 2010; Fisher 1943; Lock 1992; Poos 2004). Recent re-analysis of mortality data from 1346–1353 from Spain, Italy, England, and France suggests that plague mortalities were right around 50% (Aberth 2009). Others argue even this statistic is too low and suggest that 60% mortality was more typical (Benedictow 2004; Sloane 2013). This suggests Suffolk and Essex were not unique but rather closer to the norm. The mortalities documented in Suffolk and Essex seem to be strongly correlated with market density. Norway's non-feudal land tenure system coupled with regional and local agricultural specialization, dependent on a strong trade network appears, in a similar manner to Suffolk and Essex, to have facilitated rapid transmission of mBD across Norway (Benedictow 2004).

However, before going any further with a discussion of mBD mortalities, data from several towns illustrate the vagrancies of historical medieval data. The initiation, termination, and mortality of the mBD epidemic in Pistoia (Italy) have been identified by many authors (e.g., Chiappelli 1887; Ziegler 1969; Killinger 2002; Byrne 2004; Christakos *et al.* 2005) with varying results. Mortalities vary among 40, 50, 66, and 70%. London is similar, with mortalities varying among 35, 40, 43, 50, and 62% (Russell 1948; Megson 1998; Cohn 2002; Benedictow 2005; Christakos *et al.* 2005). Although mortality estimates tend to vary broadly, initiation dates tend to be fairly uniform among authors for each town or city. Nevertheless, the risk of mortality was highest in the larger cities because mBD resided in these cities longer than small cities or towns (Olea and Christakos 2005). Analysis of clerical deaths in the UK during the mBD suggests that most areas suffered the mBD for 4–6 months (Wood *et al.* 2003). In contrast, some small towns, villages, and hamlets had very lethal epidemics; for instance, Cadland, England suffered from May 6 through July 25, 1348 with 100% mortality, while Titchfield lost 432 of 600 residents between October 20, 1348 and May 8, 1349 (Watt 1998). In Italy, Trento lost 4,150 of approximately 5,000 residents between June 2 and September 1348 (Ginatempo and Sandri 1990).

Accurate mortality data are available in a few limited situations, including Givy in France (Gras 1939) and Saint-Nizier (Biraben 1975). The most complete daily record of mortality from Givy, France (Benedictow 2004; Olea and Christakos 2005; Gras 1939) supports a 4–6 month outbreak with the first death on July 28 to the last on November 15, 1348 (Gras 1939).

The first death from mBD at Givy in 1348 is identified as one death on epidemic day 1 (Figure 4.1). Thereafter, the peak in mBD deaths occurs on epidemic day 79,

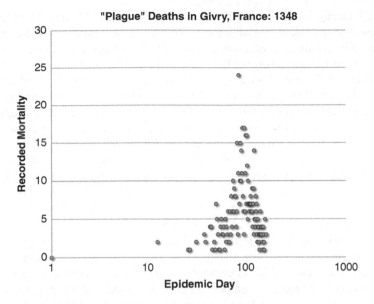

Figure 4.1 Plague deaths in Givry, France: 1348. (Data from Gras 1939.)

with 24 deaths, and the last three deaths attributed to mBD on epidemic day 149. The accuracy and length of record at Givy is unique to the 1346-1353 epidemic, in part because the village's clergyman and mortality recorder survived the plague (Gras 1939). Elsewhere, clergymen and other caregivers were especially hard hit by mBD. A record of daily mortality from St. Nizier beginning April 1, 1348 illustrates this point very graphically. Stephen Occerius died May 2, followed by his wife on May 7 and his son May 8 (Benedictow 2004). The record terminates abruptly with the death of the parish priest June 27, yet the daily number of deaths does not display a peak (Biraben 1975).

During the primary wave of mBD, infectivity was very high, perhaps approaching 100%. For about 50% of these infected, the end result was death (mortality estimates range from 30% to as high as 80% in some locations). Florence and Siena in Italy in 1348 probably suffered mortalities of 60% (Benedictow 2004, 2005). The documentary accounts of the time make clear that some of the population acquired the disease but survived. Accounts suggest that this pandemic disease was able to persist in the environment for a significant period of time. Other 'plague tracts' (written accounts of the time of the mBD) recount 'silent cities' encountered by travelers, where not only were all of the prior human inhabitants dead or fleeing, but no livestock, poultry, or domesticated animals appeared to be alive.

Gender and age mortality

The risk of mortality from mBD was not gender biased (DeWitte 2009). However, frail individuals suffered higher mortalities (DeWitte 2014; DeWitte and Wood

2008). An analysis of 207 mBD victims from East Smithfield cemetery in London suggests that the risk of mortality increased with age but mirrored normal medieval mortality (Grainger *et al.* 2008; DeWitte 2010). However, a study of 473 mBD victims from London's Royal Mint 1349 cemetery found a lower proportion of infants and 15–25 year-olds but a greater than expected representation of juveniles and 25–35 year-olds (Margerison and Knüsel 2002). In contrast, later burials at East Smithfield associated with the mBD find juvenile representation drops from 40.5% in mass burials to 29% in individual burials, whereas earlier burials associated with mBD in the southern end of East Smithfield exhibit a greater concentration of children (Grainger *et al.* 2008). Together, these East Smithfield burials suggest children and juveniles, those more susceptible, died early in the epidemic.

Throughout the medieval period, hereditary rules denied large numbers of young rural people opportunities for land ownership, thus pushing them off the land. As a consequence, 15–25 year-olds exhibited greater independence and mobility during the medieval period and comprised much of the migrant workforce that entered urban areas (Margerison and Knüsel 2002). Among 30 adult skeletons from East Smithfield cemetery, five individuals appear to be migrants with strontium and oxygen isotopes found in their tooth enamel, suggesting they had recently arrived in London from northern and western Britain (Kendall *et al.* 2013). Strontium is derived from underlying geology, while oxygen isotopes vary based on climatic factors; both are obtained through the consumption of food and water resources (Kendall *et al.* 2013). Both Boccaccio (writing during the primary wave of the medieval Black Death) and Defoe (writing later during subsequent plague epidemics) noted that "running away from it" (the plague) was the best option (Margerison and Knüsel 2002). This meant running away from the cities, and, given the migrant worker status of many 15–25 year-olds, they were best placed to do this. Juveniles and 25–35 year-olds probably remained in the towns and cities because of their investment in businesses and residences. As a result, young parents and their juvenile offspring suffered higher mortalities (Margerison and Knüsel 2002). As a rather interesting side note, the bodies placed into the mass graves in East Smithfield were not thrown in randomly. Rather, the bodies were laid out in rows; some burials had rows up to five deep, and infants and juveniles were used to fill in the spaces between the adult corpses. Some were even buried in wood boxes (Grainger *et al.* 2008) (see Figure 4.2). Across Europe during the primary wave of the mBD, rural mass burials and cemeteries catering to smaller populations, such as East Smithfield, which at the time was a community outside of the wall of London, tended to be orderly, whereas those catering to large populations exhibit randomly orientated corpses that appear to buried in a hurry (Grainger *et al.* 2008). Other mass burials exhibit a mix of both orderliness and chaos, such as those in Venice (Tran *et al.* 2011).

Birth rate

The medieval Black Death did not just kill millions, it also lowered the post 1346–1353 birth rate dramatically, with the result that Europe's population did not

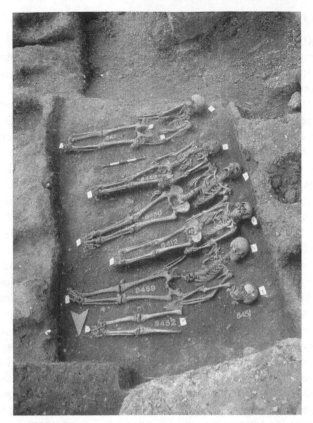

Figure 4.2 Victims of the medieval Black Death buried in the Royal Mint site at East Smithfield. (Reprinted with permission from MOLA.)

achieve pre-1346 levels until the mid to late 1500s (Bailey 1996). The shortage of labor in the post-primary wave of mBD offered women more employment opportunities, resulting in women delaying marriage and having fewer children (Poos 2004; Bailey 1996; Kitsikopoulos 2002). This demographic change, partially assisted by a general relaxing of feudal burdens, had far-reaching consequences for European socio-economic conditions, freeing money for investment capital, bringing a talented and hardworking population into the skilled labor force, and improving standards of living (Kitsikopoulos 2002).

The shortage of labor also caused wage inflation, which meant the real cost of food decreased in the post mBD environment (DeWitte 2014). Medieval Black Death survivors were then able to consume more high-quality food and, as a result, people lived longer (DeWitte 2014). In other words, some elements of the pre-mBD society benefitted from the mBD pandemic, as the primary wave of the mBD killed the weak, the elderly, and the young. The strong and fit survived in greater numbers due to the ample availability of rich food and better-paid employment opportunities (DeWitte 2014; DeWitte and Wood 2008).

Seasonality of mBD presentation

The seasonality of mBD mortality can be explained by travel to markets and fairs and pilgrimage to religious shrines. In the Middle Ages, most commerce would take place during the warmer months of the year. The seasonal nature of commerce mirrored sharp increases in peak mortality in the mBD from 1347–1353 in the spring, summer, and early fall seasons. Florentine merchant Giovanni Morelli noted that the peak of mBD mortality in 1348 and five subsequent mBD outbreaks occurred in June or July (Branca 1986; Cohn 2008). Daily aggregated Black Death records from Givry, France in 1348 (probably the most complete daily dataset of mortality during the primary wave: Gras 1939), St. Nizier-de-Lyon in 1348 (a partial record of the primary epidemic wave: Biraben 1975), central England (Halesowen: Razi 1980), the Diocese of York (actual clerical deaths from York, Cleveland, East Riding, Nottingham: Thompson 1911), and Wales (deaths in Colyan, Llanfair, Dogg, Abergwillar, Rutin Town: Shrewsbury 1971) all support this contention. Between 1348 and 1654, Norway never experienced a winter outbreak of mBD (Benedictow 2004, 2005). These data support the notion that peaks in plague transmissivity and mortality occurred when populations would be outside working the fields, attending markets and fairs, or traveling for commerce or for religious purposes (Welford and Bossak 2009).

According to Welford and Bossak (2010), inverse seasonality of infection and mortality seems logical considering an approximately 32–37 day incubation period, as proposed by Scott and Duncan (2004). In this amount of time, infectious (but not yet symptomatic) merchants and traders could travel far enough to spread the disease before becoming seriously ill, resulting in consonance between the peak time frame for trade and the peak of mBD in the late spring and summer (Bossak and Welford 2009). Some (e.g., Christakos *et al.* 2007) have suggested that this could be due to climate changes in Europe at the time of the mBD. However, the answer may be even simpler. Plague was transmitted to travelers, religious pilgrims, and fairgoers traveling along established trade routes, as has been suggested by others including Scott and Duncan (2004). As a result, the proximity to trade routes controlled the severity of mBD mortality. An examination of mortality indicates a cline with higher mortality closer to trade routes and lower mortalities with distance and isolation from established routes (e.g., roads, Bossak and Welford 2015; rivers, Yue *et al.* 2016). Furthermore, when one considers the timing during the year when medieval travelers would begin their pilgrimage to markets or religious shrines, it becomes possible to explain the observed seasonality in mortality. The peak of mortality in summer was due to the greater number of travelers coming in contact with each other along trade routes, and at markets, which were most frequent during warmer weather. As the weather began to warm, perhaps in late February or early March, travelers would begin to take to trade routes to go to markets, fairs, or religious sites. Therefore, disease numbers would start rising about 30 days later, or late March to early April, with a peak in mortality in June, July, and early August. As the temperatures started cooling again in the fall, markets and fairs would be occurring less frequently, probably dropping off considerably after October. Therefore, mortality rates from the mBD would drop in November until about February of the next year, when the cycle would pick up again. Thus seasonal changes may have determined the timing of mBD transmission, thereby providing one possible explanation for document-supported mortality peaks during the months of greatest human commingling and trade intercourse in medieval Europe.

It should also not be forgotten that the seasonality exhibited by the primary wave of the mBD and subsequent mBD pandemics that raged across Europe, with mortality peaks in June through August, flies in the face of our understanding on flea-vector dynamics. Bubonic plague pandemics, as clearly witnessed in China and India, were sustained when temperatures remained below 27.5°C and when humidity was high (Cohn 2008; Perry and Fetherston 1997; Gage and Kosoy 2005). Today, bubonic plague epidemics occur in late fall, winter, and early spring. Above 27.5°C, *Yersinia pestis* blocks flea esophagi, thus inhibiting plague transmission when fleas attempt to feed (Cavanaugh 1971). In India and China, the late fall, winter, and early spring are times when people stay indoors, thus increasing their probability of exposure to rat fleas. Today in Madagascar, bubonic plague incidence exhibits seasonal peaks, which correspond to periods when endemic rats (that play host to plague-susceptible fleas) occupy human habitation when forage in the surrounding areas is limited (Rahelinirina *et al.* 2010). This suggests that the coincidence of low forage for rats, higher rat flea activity, and humans retreating to heated buildings (that remain more humid than the outdoors) between late fall and early spring seed modern plague epidemics. In contrast, between 1346 and 1879, the mBD continued to erupt and exhibit peak mortality in the Mediterranean area during the warm-hot and dry summer months of June and July (Cohn 2008).

Transmission of plague by the human body louse (*Pediculus humanus*) has also been postulated as the vector for mBD (Ayyadurai *et al.* 2010; Drancourt *et al.* 2006; Houhamdi *et al.* 2006; Hufthammer and Walløe 2013), but here again louse vector dynamics do not offer the means to explain summer peaks in the mBD mortality observed between 1346 and 1879. Certainly experimental and archaeological work suggests human lice can transmit plague, particularly in cold weather (Ayyadurai *et al.* 2010; Houhamdi *et al.* 2006) and among populations living in poor-hygienic conditions during times of war and social disruption such as among those Napoleonic troops retreating from Russia in the 1812–13 winter (Raoult *et al.* 2006), or among the destitute and homeless (Badiaga and Brouqui 2012). Only one outbreak, a familial plague outbreak in Morocco, has been attributed to human lice (Drancourt *et al.* 2006). The question, then, is: Why are so few modern plague epidemics associated with human lice? Simply put, human lice lack the mobility to move quickly among different human hosts. In fact, inter-individual transmission of lice necessitates physical contact between an infected and uninfected human (Bechah *et al.* 2008). However, disease transmission from louse to human occurs through contamination of bite sites, conjunctivae, and mucous membranes with infected feces or crushed lice bodies (Bechah *et al.* 2008). Today, infestations of lice are found only in cold weather environments, poverty-stricken communities, and among the homeless (Bechah *et al.* 2008).

Social/cultural/economic/political ramifications

The fear and hysteria generated during the medieval Black Death had widespread and divergent consequences for art, religion, education, diet, landownership and migration, wealth inheritance, wages, employment and employment laws, geopolitical influence, minorities, warfare, and community survival.

Art

Art in Italy in 1350 underwent an abrupt transition; many of the fundamental elements and traditions were rejected and novel elements created (Meiss 1951). Images of death, despair, and destruction became widespread, with Hieronymus Bosch and Pieter Bruegel the Elder offering some of the most famous examples (DesOrmeaux 2007). The arrow, representing divine punishment, became a ubiquitous artistic symbol during and immediately after 1346–1353 (DesOrmeaux 2007). Meiss argues that art after the medieval Black Death was "pervaded by a profound pessimism" (Meiss 1951). But despair and pessimism were not the only reactions; many paintings from the time illustrate increased religiosity (Meiss 1951). The individuation of graves and burial places also appears at this time, a possible precursor to the Enlightenment, although in Florence, grave markers continued to identify ancestral lineage (Cohn 2000).

Fashion

Fashion also seems to have changed; prior to 1340, the clothes of men and women were quite similar except for the head covering, shape of neckline and length of skirt; thereafter clothing began to diverge into gendered styles (Newton 1980).

Diet

Tied to the previously noted changes, agricultural practices also changed in the post-primary wave of mBD. In several places, particularly England, nobles and freeholders moved away from labor-intensive grain cultivation practices to less labor-intensive arable production of sheep for wool and meat, and cattle for milk and meat (Gottfried 1993; Routt 2008). Many freeholders and some nobles also became more attuned to urban demands and began to cultivate a variety of cash-crops, including apples, vegetables, hops and flax, among other crops (Gottfried 1993; Routt 2008). The expansion of the commercial rabbiting using rabbit warrens in England illustrates this change toward cash-cropping and freeholder farming in the post-primary mBD environment (Bailey 1988). After 1350, rabbit-warren management became a highly lucrative source of income and a major source of red meat (Bailey 1988). Initially managed by landlords, these warrens were leased to rent-paying farmers in large numbers post-1350 as part of a larger transition to rentier farming (Bailey 1988). However, these changes in arable practice did not translate uniformly to improvements in diet and health (Yoder 2006). Bioarchaeological analyses of skeletal material from post-mBD individuals in Denmark suggest that rural peoples maintained a poor quality, grain-based diet and exhibited poorer health (Yoder 2006). In contrast, urban elites had a more marine-based diet and better health (Yoder 2006).

Education

Suffering from a paucity of students, five of the 30 European universities failed, yet Cambridge University saw the creation of four new colleges between 1348 and 1362, while Oxford created two new colleges before 1372 (Herlihy 1997). Elsewhere, the University of Florence was created in 1350. Ultimately the old order of inward-looking universities dissolved, and new ideas and curriculums, combined with knowledge from the Middle East obtained through the Crusades, moved scholars toward the Renaissance (Herlihy 1997).

Medicine

By 1348, physicians such as Gentile da Foligno, Mariano di Ser Jacopo, and Matteo Villani, and many chroniclers across Europe no longer doubted that plague was a contagious disease, but how it was transmitted was still open to question (Cohn 2002; Byrne 2006). It was only Villani who suggested that touching others transmitted the contagion (Byrne 2006; Villani *et al.* 1979). Nonetheless, physicians were recommending that their clients avoid mixing with others at markets, fairs, church congregations, taverns, and even at funerals (Cohn 2002). As a consequence, many people, but especially the rich and middle class, fled congested urban areas for the relative isolation of rural areas. Others isolated themselves from the masses, perhaps the most famous being Pope Clement VI, who restricted himself to the papal seat in Avignon. Doctors and chroniclers also began to note plague clusters among families (Cohn 2002), and by the middle of the 15th century, doctors were beginning to recognize the randomness of plague infection, and the acquisition of immunity, especially among suckling infants of mothers who survived the plague, and gravediggers (Cohn 2002).

Although these insights are especially noteworthy, medicine did not really change. Physicians still resorted to ideas and texts passed down from either Aristotle or the Hippocratic-Galenic tradition (Cohn 2002). Some doctors, e.g., Guy de Chauliac, Agnolo di Tura, and Atteo Villani, were pragmatic enough to acknowledge they could offer nothing to help their patients (Cohn 2002). A rudimentary understanding of disease spreading developed, however, and quarantine was employed (first at Pistoia, Italy in 1348 and then in Milan, Vienna and the Balearic Islands, Spain) to reduce disease risk to those community members who had not succumbed to the plague. But it was only in the post 1347–1352 environment that we saw doctors begin to develop a real appreciation for the disease, its spread, and its eradication.

Environmental changes

During the 1200s CE, global increases in CO_2 documented in ice-core data mirror widespread and sustained forest clear-cutting in western Europe, as agricultural activities expanded up to, including, and beyond the peak of the medieval warm period (van Hoof *et al.* 2008). Evidence from pollen and stomatal frequency analysis of preserved land plants suggests that forest regrowth and hence carbon sequestering following the primary wave of the medieval Black Death helped terminate the Medieval Warm Period and introduce the Little Ice Age (LIA: 1300–1900 CE) (Ruddiman 2003; Eiríksson *et al.* 2006; van Hoof *et al.* 2006). In Britain and France, although cereal production was in decline before 1347–1352, abandoned cereal-crop land underwent vegetation succession, with woody tree species flourishing in the post-primary wave environment (Yeloff and Van Geel 2007). In southern Europe, pollen typical of a mixed deciduous forest reached a peak in 1550 some 100 years after a similar peak in northern and western Europe (Mensing *et al.* 2013). However, modeling plague-induced carbon storage in Europe on land seems to only account for a 2.0 ppmv decrease in CO_2 (Ruddiman 2007).

Post-contact (with Europeans) indigenous population collapse, fire reduction (reducing CO_2 release), and reforestation (sequestering CO_2) in the Americas also appear to be significant forcing mechanisms of Little Ice Age cooling (Nevle and Bird 2008; Dull *et al.* 2010). Pandemic smallpox, measles, and influenza, warfare, enslavement, and fertility decline reduced pre-contact American populations by as much as 90%, with as many as 36–90 million indigenous peoples dying between 1500–1700 (Nevle and Bird 2008). However, the demographic-induced collapse of Pre-Columbian farmers between 1500 and 1750 appears to only account for an additional 2–5 ppmv decrease in CO_2 (Dull *et al.* 2010). Elsewhere another ~10 ppmv decrease in atmospheric CO_2 has been associated with increased Antarctic bottom water between 1550 CE and 1800 CE (Bertler *et al.* 2011).

More recently, fluctuations in global temperatures of an amplitude to initiate the Little Ice Age (LIA) have been associated with the North Atlantic conveyor and the North Atlantic Oscillation (NAO) (Alley 2007; Broecker *et al.* 1985; Larsen *et al.* 2013; Trouet *et al.* 2009; van Hoof *et al.* 2008). The North Atlantic conveyor is part of global thermohaline circulation (THC) driven by the difference in surface and deep-ocean water temperatures and saline gradients. The NAO is the difference in pressure at sea level between the Icelandic low and Azores high pressure. Coupled ocean-atmospheric general circulation models suggest an increase in mid-latitude blocking anticyclones over Europe during the LIA and a negative NAO decrease in the frequency but not intensity of mid-latitude cyclones that themselves move moisture further south into the Mediterranean and northern Sahara, leaving northern Europe cold and dry (Trouet *et al.* 2012). It is quite conceivable that reforestation and carbon sequestering during the medieval Black Death and the immediate post-1492 environment in the Americas enhanced changes in THC and NAO, but it is generally accepted that LIA was triggered by the 1259 CE volcanic eruption (Sicre *et al.* 2011).

Gender

Changes to female working and ownership conditions were also observed following the mBD. In an inquest in Wiltshire (1352–53) in the Highworth and Cricklade Hundreds, 71 individuals were taking excessive wages (that is wages above the new legal limits) of which 15 were women (Penn 1987). Among these women, eight were brewers, two peddlers and three field workers, while in the Chippenham Hundred, 51 women out of 101 individuals were taking excessive wages (Penn 1987). Many rural women in the post-primary wave mBD developed their own careers and incomes and many supplemented artisan employment as weavers and spinners with seasonal paid farm work. Others inherited landholding when their husbands died or where there were no surviving sons (Mate 1998; Penn 1987). However, these gains were temporary (Hatcher 1994). By the 1500s in Sussex and elsewhere in England, men once more dominated high-status, well-paid, supervisory employment, and brewery ownership (where some gender parity had been achieved) had returned to pre-plague norms (Mate 1998).

Nevertheless, the increased job opportunities afforded women immediately after the primary wave of the mBD in northern Europe, appear to have led women to marry later, leading to a decline in fertility (Bailey 1996; Poos 2004; Pamuk 2007). Although not mirrored in Italy (Pamuk 2007), this difference might explain why populations in northern Europe, in particular the UK, took longer to achieve pre-mBD population levels in contrast to southern Europe, while real higher wages in northern Europe were also sustained for a longer period (Pamuk 2007).

In fact, across England the number of females (be they daughters, collaterals, or male collaterals descended from the female line) inheriting land rose from 14% to 36.8% immediately following the primary wave of the mBD (Payling 1992). Among the aristocracy in Nottinghamshire from 1346–1428, 65% of land transfers were from one family to another through female inheritance; this is a level of transfer previously observed over five generations (Payling 1992). Land speculation by cash-strapped landowners also decreased in the immediate post-1349 period; this opened up opportunities for new freeholders, new female landowners, and the newly rich mercantile group to purchase land for upward social mobility not for economic gain (Payling 1992). So rather than attributing social mobility into and within the English landed gentry to the Dissolution of the Monasteries (1538 and 1541), it seems more clearly linked to the catastrophic effects of mBD (Payling 1992).

Geopolitical influence

The beginning of the decline of the Mediterranean as a political, social and economic power, to be replaced by England and the Netherlands, has been dated to the primary wave of the mBD (Gyug 1983; Vilar 1962; Vives 1955). For instance, in Barcelona, every bank except one failed between 1381 and 1383, while Cataluña peasants were subject to strict enforcement of seigneurial (feudal) land tenure systems through to the late 1400s (Gyug 1983). At the same time that power was waning in Mediterranean Europe, with the failure of the Genoese commercial trading giant following the primary wave of mBD, the Ottoman state expanded to become, within 30 years of 1352, the Ottoman Empire with vassal states across the eastern Mediterranean and Near East (Schamiloglu 2004; Varlik 2008).

The political and social decline of Mediterranean Europe coincides with the growth in the wage gap between north-west Europe and the Mediterranean. This wage gap developed as north-west Europe was able to increase productivity in agriculture, manufacturing, trade, and technology, yet it was prior to rising urbanization and the new trade that developed with Asia and the Americas in the late 1400s (Pamuk 2007). Land tenure practices, in particular the expansion of freeholding after 1347–1353, led to significant and sustained increases in agricultural productivity, as free farmers were able to follow urban market trends in consumption (Bailey 1988; Gottfried 1993; Pamuk 2007; Routt 2008). In the post 1348–1353 environment, rigid oversight and control of guilds was relaxed, interest rates declined, trade skills became a premium, rural industries expanded, and labor-saving innovations were supported (Pamuk 2007). Together these factors contributed to increased urbanization in north-western Europe and the area economically surging ahead on the Mediterranean.

Landownership and migration

Millions fled before the mBD, abandoning villages (known thereafter as plague villages) and even towns (Thompson 1921). Many of the old noble lineages died out, and with them the codes of chivalry, to be replaced by a new, vulgar *noblesse*, created as favors by surviving kings (Thompson 1921). However, in Siena, where the plague killed upwards of 50% of the population, the post-mBD government remained in the hands of the same rich elite families who ran Siena prior to the crisis (Bowsky 1964).

Minorities

Certainly, the mBD offered many opportunities to further (Other) minorities, for instance Catalans in Sicily (Cohn 2007b, 2012), beggars, clergymen, Jews, and lepers by blaming them for the catastrophe. Between 1347 and 1353, mobs, motivated by greed, Christian leadership and fear, attacked Jewish neighborhoods, killing hundreds if not thousands, and forced many surviving Jews into exile (Foa 2000). In the spring of 1348, one of the first Jewish pogroms was ignited by physicians in Languedoc claiming that Jews poisoned the air, water and food (Aberth 2005). A communication between officials in Nabonne and Girona dated April 17, 1348, stated that "enemies of the kingdom of France" – this is code for Jews – had paid several destitute men to poison wells (Aberth 2005). The first direct accusation of well-poisoning by Jews occurred in the fall of 1348 in Switzerland. By early 1351 these accusations were commonplace events in the lives of Jews throughout Europe but in particular in Germany (Aberth 2005). For instance, in Savoy in September and October of 1348, ten Jews were charged

Figure 4.3 Anonymous medieval drawing of Jews being burnt to death during the medieval Black Death (circa 1375). (Open source.)

and executed for well-poisoning (Aberth 2005). Among the worst atrocities were the several hundred Jews burnt to death on February 14, 1349 in Strasbourg (Figure 4.3).

In fact, by 1351, in excess of 350 separate massacres had been carried out and some 210 Jewish communities exterminated (Foa 2000; Ligon 2006). Various flagellant mobs, comprising artisans and peasants, burned Jewish property and killed Jews in what has been described as a 'class struggle' against an elite protected class (Jews) across German-speaking Europe, France, Spain and the Low Countries; only Italy escaped such horrors, but Mantua and Parma suffered Jewish pogroms (Cohn 2007b). These pogroms fundamentally changed Jewish civilization and left a sociological legacy that still affects Jewish communities to this day.

Incredibly, persecution of Jews in the primary wave of the mBD in German towns strongly correlates with anti-Semitic violence in the 1920s, votes for the Nazi Party before 1930, anti-Semitic letters to newspapers, organized deportations, and attacks on synagogues during the 'Night of Broken Glass' in 1938 (Voigtlaender and Voth 2011). The Hanseatic cities of the north and cities south of Cologne, but only those that encouraged immigration, do not follow this ugly trend – instead their openness to trade and migration undermines persistent racial hatred for Jews in other areas (Voigtlaender and Voth 2011).

Jews were persecuted during the time of the mBD in part because many seemed immune to the disease. Jews were known for their trading range and association with merchants from other regions and lands during medieval times. It is possible that Jewish merchants would have been exposed earlier to mBD than the general population in Europe due to trade and mercantilism. Given a restricted gene pool dictated by Jewish custom, resistance to such a mBD could have been passed down in a generation or two before 1347.

Their social exclusion from mainstream society through their restriction to Jewish ghettos might also have worked in the Jews' favor, reducing their exposure to human-human pandemics among the general population. But the spatial restriction of Jews to ghettos facilitated mass anti-Semitic violence in the form of pogroms during the primary wave of the medieval Black Death, as Christians blamed Jews for the pandemic and the poisoning of wells and food (Cohn 2012). In fact, more than a thousand Jewish communities were eradicated, with their members rounded up and slaughtered and synagogues burned (Cohn 2012). After the primary wave of the medieval Black Death (1346–1351), Jews were rarely collectively blamed for the re-emergence of the plague, but Jewish second-hand goods traders breaking quarantine laws were targeted in the late 16th and early 17th centuries in Italy (Ortensi 1579; Cohn 2012).

Religion

The failure of the clergy to save people had several divergent impacts. Across much of Europe, the mBD strengthened Christian religious beliefs leading to the creation of many new 'plague saints', religious orders, and shrines (Ligon 2006), while elsewhere the mBD weakened religious beliefs (Sherman 1991). Many Europeans thought Judgment Day had arrived; others suggested that mBD heralded the arrival of the Antichrist (Learner 1981). In this context, the flagellants were a good contemporaneous example of the chaotic, and even anarchist, response to the fear generated by mBD (DesOrmeaux 2007). However, the flagellants and new saints, religious orders, and shrines all served to criticize the prevailing religious leadership and their failure to save citizens, and this ultimately undermined religious orthodoxy and European culture (Herlihy 1997). Certainly, the failure of medical practices used by religious orders (bleeding, veneration of saints and saints' shrines) led to a new interest in herbal and practical medicinal treatments (Cantor 2001; Herlihy 1997; Twigg 1989).

Although clergy in Cambridgeshire suffered in excess of 50% mortality, bureaucratic life went on; church patrons continued to recommend candidates for vacant church offices, and bishops' manors rapidly recovered their lost rental and serf-supported agricultural production (Aberth 1995). However, many untrained, poorly trained, or dishonest and frequently incompetent clergy were ordained. Something similar happened within the bureaucracy of many if not all European governments, resulting in entrenched cronyism, corruption and subsequent mismanagement (Thompson 1921).

Taxation

The timing of the medieval Black Death was bad for all of Europe but particularly so for France, which was beset with limited finances following a war it had just lost (Henneman 1968). In the immediate aftermath of the plague in 1351, the king of France enacted a sales tax on February 24, 1351 to secure and stabilize government funding in a post-plague environment (Henneman 1968). Collections were met with opposition, and in mid-May military summons were issued to collect taxes (Henneman 1968). For example, in the south of France, Languedoc, with its weak military position located adjacent to English Gascony, although very hard hit by the plague, contributed four times more money than in 1350 to the government (Henneman 1968).

Wages and employment

Contrary to popular belief, real wages (where real wages = nominal wage index/ consumer price index) in the post-primary wave of the mBD did not rise (Hatcher 1994; Munro 2004). In fact, real wages did not achieve the previous peaks of 1336–40 until the late 1370s (Munro 2004). Certainly, the acute labor shortage following 1348–49 in England saw rural peasants gain only marginal improvements in wages; however, wage rises were the exception in London (Hatcher 1994). Nevertheless, short-lived spikes in wages did occur immediately after the mBD. In Fordham, All Saints (Suffolk, England), wages for harvest reaping rose 67% (West Suffolk Record Office source 3/15.7/2.4). While in Cuxham (Oxfordshire) wages tripled (Farmer 1991; Routt 2008). The medieval Black Death saw conservative European monarchs legislate against the survivors in an effort to control labor and artisan prices (Cohn 2007a). In England, as well as Portugal and Central Europe, labor laws, such as England's Ordinance of Labourers (1349) and Statute of Labourers (1351) were implemented to favor the interests of the aristocracy by capping prices, and restricting labor mobility and wages (Lis and Soly 2012; Penn 1987). The goal of these draconian measures was to ensure that land owned by the aristocracy was ploughed, sown, and harvested by the remaining serfs at wages and conditions that had been set pre-1348. But with labor in short supply, many serfs sought to better their conditions by fleeing to places where wages were higher or where land was available to rent (Penn and Dyer 1990). For example, in Redgrave (Suffolk) the loss of manor workers due to mBD in 1349–50 was followed by an equally damaging abandonment of manor-holding by surviving tenants in 1350–51 (Routt 2008). Durham Priory, one of the biggest landowners in England, responded very aggressively after it lost more than 50% of its tenants (Britnell 1990). The Bishop of Durham, who had palatinate powers, employed local government officers – coroners – to recapture all those tenants who fled their lands seeking higher wage employment elsewhere, and the bishop also confiscated land from anybody incapable of tilling the land and paying rent (Britnell 1990). It is important to note that the Palatinate of Durham was unusual for the time, as much of its 37,000 acres of land and 137 villages was contiguous and easily marshaled. Nevertheless, the futility of these draconian measures is evident in France where in 1351 the French government rescinded measures created in 1349, allowing wages to increase by one third (Routt 2008).

Town and city councils in Italy responded in quite different ways to the loss of citizens and the increased cost of labor. The *Popolo* and *Comune*, the two principal legislative bodies in Florence, did not try to limit or reduce urban work force wages, although these councils did try to control rural wages among their sovereign territories (Cohn 2007b). The Florentine wool guild, Arte della Lana, actually revised salaries upward in the immediate post Black Death period (Cohn 2007b). In July 1349, Florence, concerned about security, increased the salaries of its foot soldiers, crossbowmen and constables but decreased the number of police constables. Brugge in Belgium also increased police salaries (Cohn 2007b). In contrast, Orvieto (in June 1348) and Siena (in 1349) offered tax exemptions to encourage people to repopulate their towns, no matter what their trade skills (Cohn 2007b).

Beyond the immediate post-primary wave environment of the mBD (1347–1353), real wages did rise and did not return to pre-mBD levels until the 16th century, at the same time that populations returned to pre-mBD levels (Pamuk 2007). This suggests that rural depopulation during the mBD allowed rural laborers to demand and receive higher wages, but once rural populations returned to pre-mBD levels, these economic conditions ceased to exist. Higher real wages coupled with scarce labor also triggered the development of many labor-saving agricultural innovations, which ultimately led to the commercialization of agriculture and increased urbanization rates in northern Europe such that by 1600 the percentage of urban dwellers in northern Europe matched southern Europe (Pamuk 2007).

Nevertheless, government, judicial, and feudal records from the time, written typically by wealthy landowners, portray peasants as selfish, greedy, and lazy (Hatcher 1994). Feudal records from the period note workmen requesting daily contract hire rather than lengthier work (Hatcher 1994). English Law and governance responded in several ways to the demographic catastrophe that hit England in 1348–49 (Palmer 2001). Initially, the government tried to maintain the pre-1348 feudal employment status quo through several labor ordinances and statutes (Lis and Soly 2012; Palmer 2001; Penn 1987). Thereafter, the king's government began to see itself as directly responsible for society, issuing writs to ensure contracts were observed, rather than exercising indirect control through feudal lords (Musson 2000; Palmer 2001). Musson (2000) contends that these changes really began in the 1370s after several additional mBD epidemics and famines. Nevertheless, the medieval Black Death does seem to have galvanized the king's government between 1348 and the 1370s to create and maintain an English state where law was used to create, implement, and sustain social policy.

Evidence from rural Walsham in Suffolk, which suffered between 45% and 55% mBD-induced mortality, suggests that the socio-economic impacts were muted because of the lower than average mortality among those working the land (Lock 1992). However, Walsham witnessed a mass refusal to harvest manor land in 1353 – something unexpected of feudal serf peasants (Lock 1992). This is some of the first evidence of feudal peasants acting collectively to disobey their manor overlords. Poos (2004) estimates that in 1377 one in seven Essex rural working individuals were fined for violating labor legislation. These local civil disobediences manifested themselves in more widespread rebellions in the French Jacquerie of 1358 and the English Peasants' Rebellion of 1381 (Hilton 2003; Poos 2004). At the heart of these uprisings was the attempt by serfs to control their own destiny. Prior to the medieval Black Death, many nobles had freed serfs (at a price) to become rent-paying tenants on their land. Nobles quickly perceived that these rent-paying tenants were more productive, and so this trend was encouraged. However, the post medieval Black Death labor shortage saw nobles try to stop and even reverse this trend through the Ordinance of Laborers enacted in England in June 1349, but both the rebellions and nobles' attempts to once more exclusively control rural labor failed (Hatcher 1994; Poos 2004).

Such broad statements hide the spatial variation and economic complexity of the peasant-nobility dynamics in a post-1348 environment. For instance, in England, Kent prospered while adjoining Sussex suffered long-term economic depression (Walmsley 1992). In Kent, greater divisions existed between rich and poor than in Sussex, which fostered a more mobile labor force; in response, Kentish landlords were forced to exercise less feudal power and offer higher wages to maintain labor, and so by the late 1400s, Kent was four times wealthier (Walmsley 1992). Ultimately, the labor shortages immediately following the primary wave of mBD began to sever the ties of feudal bondage across Europe (Platt 1997; Haddock and Kiesling 2002); in particular, it appears that the mBD and the Peasants' Revolt combined to end excessive taxation of the poor (Schofield 2009).

Warfare

Medieval warfare was also affected. In France, military service was a major component of feudal obligation in the pre-1348 system. Following the primary wave of mBD, the French government saw tax revenues decline sharply yet wages of mercenary soldiers rise, and fewer feudal soldiers to fill in the gaps. As a result, the French copied the English by instigating a lower-paid, professional army (Huppert 1998).

Summary

Simply put, the primary wave of the medieval Black Death changed nearly everything. It was for all intents and purposes the beginning of the end for feudalism and the beginning of agricultural wage-labor. It introduced quarantine practices through the implementation of public health policies and codes, it forced countries to transition from unpaid feudal-obligated armies to professional armies, it began the slow transition from a southern-dominated Europe to a northern-dominated Europe, it undermined religious orthodoxy, it changed art, and even changed fashion, education, and language. In England, for example, the aristocrats, clergy, and playwrights began to transition from speaking Norman French to early English. Sadly, it was a false dawn for female rights.

References

Aberth, J., 1995. The Black Death in the diocese of Ely: the evidence of the bishop's register. *Journal of Medieval History*, *21*(3), pp.275–287.

Aberth, J., 2005. *The Black Death: the great mortality of 1348–1350* (p. 200). London: Palgrave Macmillan.

Aberth, J., 2009. *From the brink of the Apocalypse. Confronting famine, war, plague, and death in the later Middle Ages*. London: Routledge.

Adjemian, J.Z., Foley, P., Gage, K.L. and Foley, J.E., 2007. Initiation and spread of traveling waves of plague, Yersinia pestis, in the western United States. *The American Journal of Tropical Medicine and Hygiene*, *76*(2), pp.365–375.

Alexander, J.T., 2002. *Bubonic plague in early modern Russia: public health and urban disaster*. Oxford: Oxford University Press.

Alley, R.B., 2007. Wally was right: predictive ability of the North Atlantic "conveyor belt" hypothesis for abrupt climate change. *Annual Review of Earth and Planetary Sciences*, *35*, pp.241–272.

Amori, G. and Cristaldi, M., 1999. Rattus norvegicus. *The Atlas of European Mammals*, pp.278–279.

Arandez, A.S., 1980. La reconquista de las vías marítimas. *Anuario de Estudios Medievales*, *10*, p.41.

Ayyadurai, S., Sebbane, F., Raoult, D. and Drancourt, M., 2010. Body lice, Yersinia pestis orientalis, and Black Death. *Emerging Infectious Diseases*, *16*(5), p.892.

Badiaga, S. and Brouqui, P., 2012. Human louse-transmitted infectious diseases. *Clinical Microbiology and Infection*, *18*(4), pp.332–337.

Bailey, M., 1988. The rabbit and the medieval East Anglian economy. *The Agricultural History Review*, *36*, pp.1–20.

Bailey, M., 1996. Demographic decline in late medieval England: some thoughts on recent research. *The Economic History Review*, *49*(1), pp.1–19.

Bailey, M., 2010. *Medieval Suffolk: an economic and social history, 1200–1500 (Volume 1)*. Suffolk, UK: Boydell & Brewer.

Bechah, Y., Capo, C., Mege, J.L. and Raoult, D., 2008. Epidemic typhus. *The Lancet Infectious Diseases*, *8*(7), pp.417–426.

Benedict, C.A., 1996. *Bubonic plague in nineteenth-century China*. Palo Alto, CA: Stanford University Press.

Benedictow, O.J., 2004. *The Black Death, 1346–1353: the complete history*. Suffolk, UK: Boydell & Brewer.

Benedictow, O.J., 2005. The Black Death: the greatest catastrophe ever. *History Today*, *55*(3), p.42.

Berthe, M., 1984. *Famines et épidémies dans les campagnes navarraises à la fin du Moyen-Age*. Paris: S.F.I.E.D.

Bertler, N.A.N., Mayewski, P.A. and Carter, L., 2011. Cold conditions in Antarctica during the little ice age – implications for abrupt climate change mechanisms. *Earth and Planetary Science Letters*, *308*(1), pp.41–51.

Biraben, J-N., 1975. *Les hommes et al. peste en France et dans les pays europeens et mediterraneens*. Volumes 1 (455 p.) and 2 (416 p.). Paris, France: Mouton.

Boots, M. and Mealor, M., 2007. Local interactions select for lower pathogen infectivity. *Science*, *315*(5816), pp.1284–1286.

Boots, M. and Sasaki, A., 1999. 'Small worlds' and the evolution of virulence: infection occurs locally and at a distance. *Proceedings of the Royal Society of London B: Biological Sciences*, *266*(1432), pp.1933–1938.

Bos, K.I., Herbig, A., Sahl, J., Waglechner, N., Fourment, M., Forrest, S.A., Klunk, J., Schuenemann, V.J., Poinar, D., Kuch, M. and Golding, G.B., 2016. Eighteenth century Yersinia pestis genomes reveal the long-term persistence of an historical plague focus. *Elife*, *5*, p.e12994.

Bos, K.I., Schuenemann, V.J., Golding, G.B., Burbano, H.A., Waglechner, N., Coombes, B.K., McPhee, J.B., DeWitte, S.N., Meyer, M., Schmedes, S. and Wood, J., 2011. A draft genome of Yersinia pestis from victims of the Black Death. *Nature*, *478*(7370), pp.506–510.

Bossak, B.H. and Welford, M.R., 2009. Did medieval trade activity and a viral etiology control the spatial extent and seasonal distribution of Black Death mortality? *Medical Hypotheses*, *72*(6), pp.749–752.

Bossak, B.H. and Welford, M.R., 2010. Spatio-temporal attributes of pandemic and epidemic diseases. *Geography Compass*, *4*(8), pp.1084–1096.

Bossak, B.H. and Welford, M.R., 2015. Reconstructing the spatio-temporal characteristics of infectious diseases: a GIS-based case study of the Medieval Black Death. In *Spatial analysis in health geography* (pp.71–84), eds. P. Kanaroglou, E. Delmelle and A. Páez. London: Routledge.

Bowsky, W.M., 1964. The impact of the Black Death upon Sienese government and society. *Speculum*, *39*(1), pp.1–34.

Branca, V. ed., 1986. *Mercanti scrittori: ricordi nella Firenze tra Medioevo e Rinascimento*. Rusconi Libri: Santarcangelo di Romagna.

Britnell, R.H., 1990. Feudal reaction after the Black Death in the Palatinate of Durham. *Past & Present*, *128*, pp.28–47.

Britnell, R., 1994. The Black Death in English towns. *Urban History*, *21*(2), pp.195–210.

Broecker, W.S., Peteet, D.M. and Rind, D., 1985. Does the ocean–atmosphere system have more than one stable mode of operation? *Nature*, *315*(6014), pp.21–26.

Bryce, J., Boschi-Pinto, C., Shibuya, K., Black, R.E. and WHO Child Health Epidemiology Reference Group, 2005. WHO estimates of the causes of death in children. *The Lancet*, *365*(9465), pp.1147–1152.

Byrne, J.P., 2004. *The Black Death*. Westport, USA: Greenwood Publishing Group.

Byrne, J.P., 2006. *Daily life during the Black Death*. Westport, USA: Greenwood Publishing Group.

Cantor, N.F., 2001. *In the wake of the plague: the Black Death and the world it made*. New York: Simon and Schuster.

Carpentier, É., 1962. Autour de la Peste Noire: Famines et épidémies dans l'histoire du XIVe siècle. *Annales: E. S. C.*, *17*, pp.1062–1092.

Cavanaugh, D.C., 1971. Specific effect of temperature upon transmission of the plague bacillus by the oriental rat flea, Xenopsylla cheopis. *American Journal of Tropical Medicine and Hygiene*, *20*(2), pp.264–273.

Chiappelli, A. ed., 1887. Gli ordinamenti sanitari del comune di Pistoia contro la pestilenzia del 1348. *Archivio Storico Italiano*, Ser. 4, *20*, pp.8–22.

Christakos, G., Olea, R.A., Serre, M.L., Wang, L.L. and Yu, H.L., 2005. *Interdisciplinary public health reasoning and epidemic modelling: the case of Black Death* (p. 320). New York: Springer.

Christakos, G., Olea, R.A. and Yu, H.L., 2007. Recent results on the spatiotemporal modelling and comparative analysis of Black Death and bubonic plague epidemics. *Public Health, 121*(9), pp.700–720.

Cohn, S.K., 2000. The place of the dead in Flanders and Tuscany: towards a comparative history of the Black Death. In *The place of the dead: death and remembrance in late medieval and early modern Europe* (pp. 17–43), eds. B. Gordon and P. Marshall. Cambridge: Cambridge University Press.

Cohn, S.K., 2002. *The Black Death transformed: disease and culture in early Renaissance Europe*. London: Arnold.

Cohn, S.K., 2007a. After the Black Death: labour legislation and attitudes towards labour in late-medieval western Europe. *The Economic History Review, 60*(3), pp.457–485.

Cohn, S.K., 2007b. The Black Death and the burning of Jews. *Past & Present, 196*(1), pp.3–36.

Cohn Jr., S.K., 2008. Epidemiology of the Black Death and successive waves of plague. *Medical History. Supplement*, (27), p.74.

Cohn, S.K., 2012. Pandemics: waves of disease, waves of hate from the plague of Athens to AIDS. *Historical Research, 85*(230), pp.535–555.

Cui, Y., Yu, C., Yan, Y., Li, D., Li, Y., Jombart, T., Weinert, L.A., Wang, Z., Guo, Z., Xu, L. and Zhang, Y., 2013. Historical variations in mutation rate in an epidemic pathogen, Yersinia pestis. *Proceedings of the National Academy of Sciences, 110*(2), pp.577–582.

Davis, D.E., 1986. The scarcity of rats and the Black Death: an ecological history. *The Journal of Interdisciplinary History, 16*(3), pp.455–470.

DesOrmeaux, A.L., 2007. *The Black Death and its effect on fourteenth- and fifteenth-century art*. MA thesis, Baton Rouge: Louisiana State University.

DeWitte, S.N., 2009. The effect of sex on risk of mortality during the Black Death in London, AD 1349–1350. *American Journal of Physical Anthropology, 139*(2), pp.222–234.

DeWitte, S.N., 2010. Age patterns of mortality during the Black Death in London, AD 1349–1350. *Journal of Archaeological Science, 37*(12), pp.3394–3400.

DeWitte, S.N., 2014. Mortality risk and survival in the aftermath of the medieval Black Death. *PloS One, 9*(5), p.e96513.

DeWitte, S.N., 2015. Setting the stage for medieval plague: pre-Black Death trends in survival and mortality. *American Journal of Physical Anthropology, 158*(3), pp.441–451.

DeWitte, S.N. and Wood, J.W., 2008. Selectivity of Black Death mortality with respect to preexisting health. *Proceedings of the National Academy of Sciences, 105*(5), pp.1436–1441.

Drancourt, M., Houhamdi, L. and Raoult, D., 2006. Yersinia pestis as a telluric, human ectoparasite-borne organism. *The Lancet Infectious Diseases, 6*(4), pp.234–241.

Drancourt, M., Roux, V., La Vu Dang, L.T.H., Castex, D., Chenal-Francisque, V., Ogata, H., Fournier, P.E., Crubézy, E. and Raoult, D., 2004. Genotyping, Orientalis-like Yersinia pestis, and plague pandemics. *Emerging Infectious Diseases, 10*(9), p.1585.

Dubois, H., 1988. L'essor medieval. In *Histoire de la population Francaise volume 1* (pp. 207–266), ed. J. Dupaquier. Paris: Presses universitaires de France.

Dull, R.A., Nevle, R.J., Woods, W.I., Bird, D.K., Avnery, S. and Denevan, W.M., 2010. The Columbian encounter and the Little Ice Age: abrupt land use change, fire, and greenhouse forcing. *Annals of the Association of American Geographers, 100*(4), pp.755–771.

Duncan, C.J. and Scott, S., 2005. What caused the Black Death? *Postgraduate Medical Journal, 81*(955), pp.315–320.

Duncan, S.R., Scott, S. and Duncan, C.J., 2005. Reappraisal of the historical selective pressures for the CCR5-Δ32 mutation. *Journal of Medical Genetics, 42*(3), pp.205–208.

Eager, J.M., 1903. The early history of quarantine: origin of sanitary measures directed against yellow fever (Volume 12). Washington DC: US Government Printing Office.

Eiríksson, J., Bartels-Jonsdottir, H.B., Cage, A.G., Gudmundsdottir, E.R., Klitgaard-Kristensen, D., Marret, F., Rodrigues, T., Abrantes, F., Austin, W.E., Jiang, H. and Knudsen, K.L., 2006. Variability of the North Atlantic Current during the last 2000 years based on shelf bottom water and sea surface temperatures along an open ocean/shallow marine transect in western Europe. *The Holocene, 16*(7), pp.1017–1029.

Eppinger, M., Guo, Z., Sebastian, Y., Song, Y., Lindler, L.E., Yang, R. and Ravel, J., 2009. Draft genome sequences of Yersinia pestis isolates from natural foci of endemic plague in China. *Journal of Bacteriology, 191*(24), pp.7628–7629.

Fagan, B., 2001. *The little Ice Age: how climate made history 1300–1850*. New York: Basic Books.

Farmer, D.L., 1991. Prices and wages, 1350–1500. *The Agrarian History of England and Wales, 3*, pp.1348–1500.

Fisher, J.L., 1943. The Black Death in Essex. *Essex Review, 52*, pp.13–20.

Fletcher, R.A., 1984. *Saint James's catapult: the life and times of Diego Gelmírez of Santiago de Compostela* (341 p). Oxford: Clarendon Press.

Foa, A., 2000. *The Jews of Europe after the Black Death* (288 p). Berkeley: University of California Press.

Fraser, E.D.G., 2011. Can economic, land use and climatic stresses lead to famine, disease, warfare and death? Using Europe's calamitous 14th century as a parable for the modern age. *Ecological Economics, 70*(7), pp.1269–1279.

Gage, K.L. and Kosoy, M.Y., 2005. Natural history of plague: perspectives from more than a century of research. *Annual Review of Entomology, 50*, pp.505–528.

Gelting, M.H., 1991. The mountains and the plague: Maurienne, 1348. *Collegium Medievale, 1991*(1), pp.7–45.

Ginatempo, M. and Sandri, L., 1990. *L'Italia delle città. Il popolamento urbano tra medioevo e Rinascimento (sec. XIII–XVI)* [Italy of cities. The urban population between Middle Ages and Renaissance (XIII–XVI centuries)]. Florence: Le Lettere.

Gottfried, R.S., 1993. The Black Death: natural and human disaster in Medieval Europe (203 p). New York: Free Press.

Grainger, I., Hawkins, D., Cowal, L. and Mikulski, R., 2008. *The Black Death cemetery, East Smithfield, London* (Volume 43). Museum of London Archaeology Svc.

Gras, P., 1939. Le register paroissial de Givry (1334–1357) et la Peste Noire en Bourgogne. *Bibl. Ec. Chartes, 100*, pp.295–308.

Green, M.H. and Schmid, B., 2016. Plague dialogues: Monica Green and Boris Schmid on Plague phylogeny (I). https://contagions.wordpress.com/2016/06/27/plague-dialogues-monica-green-and-boris-schmid-on-plague-phylogeny-i/, accessed June 1, 2017.

Gyug, R., 1983. The effects and extent of the Black Death of 1348: new evidence for clerical mortality in Barcelona. *Mediaeval Studies, 45*, pp.385–398.

Haddock, D.D. and Kiesling, L., 2002. The Black Death and property rights. *The Journal of Legal Studies, 31*(S2), pp.S545–S587.

Haensch, S., Bianucci, R., Signoli, M., Rajerison, M., Schultz, M., Kacki, S., Vermunt, M., Weston, D.A., Hurst, D., Achtman, M. and Carniel, E., 2010. Distinct clones of Yersinia pestis caused the Black Death. *PLoS pathogens, 6*(10), p.e1001134.

Harrison, D., 2000. Stora döden. Den värsta katastrof som drabbat Europa. Stockholm: Ordfront förlag.

Hatcher, J., 1994. England in the aftermath of the Black Death. *Past & Present, 144*, pp.3–35.

Hecker, J.F.C., 1838. *The epidemics of the Middle Ages*. Philadelphia: Haswell, Barrington, and Haswell.

Herlihy, D., 1997. *The Black Death and the transformation of the west.* Boston: Harvard University Press.

Henneman, J.B., 1968. The Black Death and royal taxation in France, 1347–1351. *Speculum, 43*(3), pp.405–428.

Hilton, R., 2003. *Bond men made free: medieval peasant movements and the English rising of 1381.* London: Routledge.

Horrox, R., 1994. *The Black Death.* Manchester: Manchester University Press.

Houhamdi, L., Lepidi, H., Drancourt, M. and Raoult, D., 2006. Experimental model to evaluate the human body louse as a vector of plague. *Journal of Infectious Diseases, 194*(11), pp.1589–1596.

Hufthammer, A.K. and Walløe, L., 2013. Rats cannot have been intermediate hosts for Yersinia pestis during medieval plague epidemics in Northern Europe. *Journal of Archaeological Science, 40*(4), pp.1752–1759.

Huppert, G., 1998. *After the Black Death: a social history of early modern Europe.* Bloomington: Indiana University Press.

Ibn al Khatib 1369, 1978. *al-Ihata fi akhbar Gharnata'* (The Complete Source on the History of Granada), ed. Muhammad Abd Allah Inan. Cairo: Maktabat al-Khanji.

Karlsson, G., 1996. Plague without rats: the case of fifteenth-century Iceland. *Journal of Medieval History, 22*(3), pp.263–284.

Kelly, J., 2005. *The great mortality: an intimate history of the Black Death, the most devastating plague of all time.* New York: HarperCollins.

Kendall, E.J., Montgomery, J., Evans, J.A., Stantis, C. and Mueller, V., 2013. Mobility, mortality, and the middle ages: identification of migrant individuals in a 14th century Black Death cemetery population. *American Journal of Physical Anthropology, 150*(2), pp.210–222.

Killinger, C.L., 2002. *The history of Italy.* Santa Barbara: Greenwood.

Kitsikopoulos, H., 2002. The impact of the Black Death on peasant economy in England, 1350–1500. *Journal of Peasant Studies, 29*(2), pp.71–90.

Lamb, H.H., 2013. *Climate: present, past and future (Routledge Revivals): Volume 2: climatic history and the future.* London: Routledge.

Larsen, D.J., Miller, G.H. and Geirsdóttir, Á., 2013. Asynchronous Little Ice Age glacier fluctuations in Iceland and European Alps linked to shifts in subpolar North Atlantic circulation. *Earth and Planetary Science Letters, 380*, pp.52–59.

Learner, R.E., 1981. The Black Death and western eschatological mentalities. *American Historical Review, 86*, pp.533–552.

Ligon, B.L., 2006. Plague: a review of its history and potential as a biological weapon. *Seminars in Pediatric Infectious Diseases, 17*(3), pp.161–170.

Lis, C. and Soly, H., 2012. Labor Laws in Western Europe, 13th–16th Centuries: patterns of political and socio-economic rationality. In *Working on labor: essays in honor of Jan Lucassen* (pp.299-321), ed. M. van der Linden, Brill UK.

Little, L.K., 2006. Life and afterlife of the first plague pandemic. In *Plague and the end of antiquity: the Pandemic of 541–75* (pp. 3–32), ed. L.K. Little. Cambridge: Cambridge University Press.

Lock, R., 1992. The Black Death in Walsham-le-Willows. *Proceedings of the Suffolk Institute of Archaeology and History, 37*(4), pp.316–337.

Mackay, A., 1997. *Atlas of medieval Europe.* London: Taylor and Francis.

Mann, M.E., Zhang, Z., Rutherford, S., Bradley, R.S., Hughes, M.K., Shindell, D., Ammann, C., Faluvegi, G. and Ni, F., 2009. Global signatures and dynamical origins of the Little Ice Age and Medieval Climate Anomaly. *Science, 326*(5957), pp.1256–1260.

Margerison, B.J. and Knüsel, C.J., 2002. Paleodemographic comparison of a catastrophic and an attritional death assemblage. *American Journal of Physical Anthropology*, *119*(2), pp.134–143.

Mate, M.E., 1998. *Daughters, wives and widows after the Black Death: women in Sussex, 1350–1535*. Woodbridge: Boydell Press.

Megson, B.E., 1998. Mortality among London citizens in the Black Death. *Medieval Prosopography*, *19*, pp.125–134.

Meiss, M., 1951. *Painting in Florence and Siena after the Black Death*. New York: Princeton University Press.

Mensing, S., Tunno, I., Cifani, G., Florindo, F., Noble, P., Sagnotti, L. and Piovesan, G., 2013. Effects of human impacts and climate variation on forests: the Rieti basin since medieval time. *Annali di Botanica*, *3*, pp.121–126.

Morelli, G., Song, Y., Mazzoni, C.J., Eppinger, M., Roumagnac, P., Wagner, D.M., Feldkamp, M., Kusecek, B., Vogler, A.J., Li, Y. and Cui, Y., 2010. Phylogenetic diversity and historical patterns of pandemic spread of Yersinia pestis. *Nature Genetics*, *42*(12), pp.1140–1143.

Munro, J., 2004. Before and after the Black Death: money, prices, and wages in fourteenth-century England. Knoxville: University of Tennesse-Working Paper 24.

Musson, A., 2000. New labor laws, new remedies? Legal reaction to the Black Death crisis. In *Fourteenth century England I* (pp. 73–88), ed. N. Saul. Suffolk, UK: Boydell & Brewer.

Nevle, R.J. and Bird, D.K., 2008. Effects of syn-pandemic fire reduction and reforestation in the tropical Americas on atmospheric CO_2 during European conquest. *Palaeogeography, Palaeoclimatology, Palaeoecology*, *264*(1), pp.25–38.

Newton, S.M., 1980. *Fashion in the Age of the Black Prince: a Study of the Years 1340–1365*. Suffolk, UK: Boydell & Brewer.

Noble, J.V., 1974. Geographic and temporal development of plagues. *Nature*, *250*(5469), pp.726–729.

Oeding, P., 1990. The Black Death in Norway. *Tidsskr Nor Laegeforen,110*, pp.2204–2208.

Olea, R.A. and Christakos, G., 2005. Duration of urban mortality for the 14th-century Black Death epidemic. *Human Biology*, *77*(3), pp.291–303.

Ortensi, A.C.D., 1579. I Cinque Libri degl'Avvertimenti ordini, gride, et editti. Venice.

Palmer, R.C., 2001. *English Law in the age of the Black Death, 1348–1381: a transformation of governance and law*. Chapel Hill: University of North Carolina Press.

Pamuk, S., 2007. The Black Death and the origins of the 'great divergence' across Europe, 1300–1600. *European Review of Economic History*, *11*, pp.289–317.

Payling, S.J., 1992. Social mobility, demographic change, and landed society in Late Medieval England. *The Economic History Review (New Series)*, *45*(1), pp.51–73.

Penn, S.A.C., 1987. Female wage-earners in late fourteenth-century England. *The Agricultural History Review*, *35*(1), pp.1–14.

Penn, S.A. and Dyer, C., 1990. Wages and earnings in late medieval England: evidence from the enforcement of the labour laws. *The Economic History Review*, *43*(3), pp.356–376.

Perry, R.D. and Fetherston, J.D., 1997. Yersinia pestis—etiologic agent of plague. *Clinical Microbiology Reviews*, *10*(1), pp.35–66.

Platt, C., 1997. *King death: the Black Death and its aftermath in Late-Medieval England*. Toronto: University of Toronto Press.

Poos, L.R., 1991. *A rural society after the Black Death: Essex 1350–1525* (No. 18). Cambridge: Cambridge University Press.

Poos, L.R., 2004. *A rural society after the Black Death: Essex 1350–1525*. Cambridge: Cambridge University Press.

Rahelinirina, S., Duplantier, J.M., Ratovonjato, J., Ramilijaona, O., Ratsimba, M. and Rahalison, L., 2010. Study on the movement of Rattus rattus and evaluation of the plague dispersion in Madagascar. *Vector-Borne and Zoonotic Diseases*, *10*(1), pp.77–84.

Raoult, D., Dutour, O., Houhamdi, L., Jankauskas, R., Fournier, P.E., Ardagna, Y., Drancourt, M., Signoli, M., La, V.D., Macia, Y. and Aboudharam, G., 2006. Evidence for louse-transmitted diseases in soldiers of Napoleon's Grand Army in Vilnius. *Journal of Infectious Diseases*, *193*(1), pp.112–120.

Ratsitorahina, M., Chanteau, S., Rahalison, L., Ratsifasoamanana, L. and Boisier, P., 2000. Epidemiological and diagnostic aspects of the outbreak of pneumonic plague in Madagascar. *The Lancet*, *355*(9198), pp.111–113.

Razi, Z., 1980. *Life, marriage & death in a medieval parish*. Cambridge: Cambridge University Press.

Rezakhani, K., 2010. The road that never was: the Silk Road and Trans-Eurasian exchange. *Comparative Studies of South Asia, Africa and the Middle East*, 30(3), pp.420–433.

Rielly, K., 2010. *The black rat. Extinctions and invasions: a social history of British fauna.* Oxford, UK: Windgather Press, Oxbow Books, pp.134–145.

Roseboom, T., de Rooij, S. and Painter, R., 2006. The Dutch famine and its long-term consequences for adult health. *Early Human Development*, *82*(8), pp.485–491.

Routt, D., 2008. The economic impact of the Black Death. EH.Net Encyclopedia, ed. R. Whaples. July 20, 2008. http://eh.net/encyclopedia/article/Routt.Black.Death

Ruddiman, W.F., 2003. The anthropogenic greenhouse era began thousands of years ago. *Climatic Change*, *61*(3), pp.261–293.

Ruddiman, W.F., 2007. The early anthropogenic hypothesis: challenges and responses. *Reviews of Geophysics*, *45*(4), RG4001.

Russell, J.C., 1948. Demographic pattern in history. *Population Studies*, *1*(4), pp.388–404.

Schamiloglu, U., 2004. The rise of the Ottoman Empire: the Black Death in Medieval Anatolia and its impact on Turkish civilization. In *Views from the edge: essays in honor of Richard W. Bulliet* (pp. 255–279), eds. N. Yavari, L.G. Potter and J.-M.R. Oppenheim. New York: Columbia University Press.

Schevill, F., 1961. *History of Florence*. New York: Frederick Ungar Publishing.

Schmid, B.V., Büntgen, U., Easterday, W.R., Ginzler, C., Walløe, L., Bramanti, B. and Stenseth, N.C., 2015. Climate-driven introduction of the Black Death and successive plague reintroductions into Europe. *Proceedings of the National Academy of Sciences*, *112*(10), pp.3020–3025.

Schofield, P., 2009. Review of king death. the Black Death and its aftermath in late Medieval England, (review no. 17). www.history.ac.uk/reviews/review/17, accessed May 2016.

Scott, S. and Duncan, C.J., 2004. *Return of the Black Death: the world's greatest serial killer*. Chichester, UK: Wiley.

Sherman, D., 1991. *Western civilization: images and interpretations* (Volume 1, 3rd Edition). New York: McGraw-Hill.

Shin, S-I, Liu, Z, Otto-Bliesner, Kutzbach, J.E. and Vavrus, S.J., 2003. Southern Ocean sea-ice control of the glacial North Atlantic thermohaline circulation. *Geophysical Research Letters*, *30*(2), pp.1–4,68.

Shrewsbury, J.F.D., 1971, repr. 2005. *A history of bubonic plague in the British Isles*. Cambridge: Cambridge University Press.

Sicre, M.A., Hall, I.R., Mignot, J., Khodri, M., Ezat, U., Truong, M.X., Eiríksson, J. and Knudsen, K.L., 2011. Sea surface temperature variability in the subpolar Atlantic over the last two millennia. *Paleoceanography*, *26*(4). doi:10.1029/2011PA002169.

Slack, P., 1988. Responses to Plague in early Modern Europe: the implications of Public Health. *Social Research, 55*(9), pp.433–453.

Sloane, B., 2013. *The Black Death in London.* Stroud, UK: The History Press.

Snell, M., 2017. The arrival and spread of the Black Death through Europe. www.thoughtco. com/spread-of-the-black-death-through-europe-4123214, updated July 15, 2017.

Spyrou, M.A., Tukhbatova, R.I., Feldman, M., Drath, J., Kacki, S., de Heredia, J.B., Arnold, S., Sitdikov, A.G., Castex, D., Wahl, J. and Gazimzyanov, I.R., 2016. Historical Y. pestis genomes reveal the European Black Death as the source of ancient and modern plague pandemics. *Cell Host & Microbe, 19*(6), pp.874–881.

Stenseth, N.C., Samia, N.I., Viljugrein, H., Kausrud, K.L., Begon, M., Davis, S., Leirs, H., Dubyanskiy, V.M., Esper, J., Ageyev, V.S. and Klassovskiy, N.L., 2006. Plague dynamics are driven by climate variation. *Proceedings of the National Academy of Sciences, 103*(35), pp.13110–13115.

Stothers, R.B., 2000. Climatic and demographic consequences of the massive volcanic eruption of 1258. *Climatic Change, 45*(2), pp.361–374.

Suckow, M.A., Weisbroth, S.H. and Franklin, C.L. eds., 2005. *The laboratory rat.* San Diego: Academic Press

Thompson, A.H., 1911. Registers of John Gynewell, Bishop of Lincoln, for the Years, 1347–1350. *Archaeological Journal, 68*(1), pp.300–360.

Thompson, J.W., 1921. The aftermath of the Black Death and the aftermath of the great war. *American Journal of Sociology, 26*(5), pp.565–572.

Tran, T.N., Signoli, M., Fozzati, L., Aboudharam, G., Raoult, D. and Drancourt, M., 2011. High throughput, multiplexed pathogen detection authenticates plague waves in medieval Venice, Italy. *PLoS One, 6*(3), p.e16735.

Trouet, V., Esper, J., Graham, N.E., Baker, A., Scourse, J.D. and Frank, D.C., 2009. Persistent positive North Atlantic Oscillation mode dominated the medieval climate anomaly. *Science, 324*(5923), pp.78–80.

Trouet, V., Scourse, J.D. and Raible, C.C., 2012. North Atlantic storminess and Atlantic meridional overturning circulation during the last millennium: reconciling contradictory proxy records of NAO variability. *Global and Planetary Change, 84*, pp.48–55.

Tuchman, B.W., 1978. *A distant mirror.* New York: Ballantine Books.

Twigg, G., 1984. *The Black Death: a biological appraisal.* London: Batsford Academic & Educational.

Twigg, G.I., 1989. The Black Death in England. An epidemiological dilemma. In *Maladies et sociétés (XII e-XVIII e siècles). Colloque* (pp. 75–98).

van Hoof, T.B., Bunnik, F.P., Waucomont, J.G., Kürschner, W.M. and Visscher, H., 2006. Forest re-growth on medieval farmland after the Black Death pandemic—Implications for atmospheric CO_2 levels. *Palaeogeography, Palaeoclimatology, Palaeoecology, 237*(2), pp.396–409.

van Hoof, T.B., Wagner-Cremer, F., Kürschner, W.M. and Visscher, H., 2008. A role for atmospheric CO_2 in preindustrial climate forcing. *Proceedings of the National Academy of Sciences, 105*(41), pp.15815–15818.

Varlik, N., 2008. *Disease and empire: a history of plague epidemics in the early modern Ottoman Empire (1453–1600).* The University of Chicago.

Varlik, N., 2015. *Plague and empire in the early modern Mediterranean world.* Cambridge: Cambridge University Press.

Vilar, P., 1962. *La Catalogne dans l'Espagne moderne. Recherches sur les fondements economiques des structures nationales.* 3 vols. Paris: SEVPEN.

Villani, G., Villani, M. and Villani, F., 1979. *Cronica: con le continuazioni di Matteo e Filippo* (Volume 159). G. Einaudi.

Vives, J.V., 1955. Evolucion de la economica catalane durante la primera mitad del siglo XV. Palma: Spain.

Voigtlaender, N. and Voth, H-J., 2011. Persecution perpetuated: the medieval origins of anti-semitic violence in Nazi Germany. NBER Working Papers 17113.

Walmsley, J., 1992. A review of 'The agrarian history of England and Wales. Volume III: 1348–1500'. *Parergon, 10*(2), pp.225–228.

Watt, D.G., 1998. The Black Death in Dorset and Hampshire. *The Hatcher Review, 5*, pp.21–31.

Welford, M.R. and Bossak, B.H., 2009. Validation of inverse seasonal peak mortality in medieval plagues, including the Black Death, in comparison to modern Yersinia pestis-variant diseases. *PLoS One, 4*(12), p.e8401.

Welford, M.R. and Bossak, B.H., 2010. Body Lice, *Yersinia pestis Orientalis*, and Black Death. *Emerging Infectious Diseases, 16*(10), pp.1650–1651.

Wheelis, M., 2002. Biological Warfare at the 1346 Siege of Caffa. *Emerging Infectious Diseases, 8*(9), pp.971–975.

Wilschut, L.I., Laudisoit, A., Hughes, N.K., Addink, E.A., Jong, S.M., Heesterbeek, H.A., Reijniers, J., Eagle, S., Dubyanskiy, V.M. and Begon, M., 2015. Spatial distribution patterns of plague hosts: point pattern analysis of the burrows of great gerbils in Kazakhstan. *Journal of Biogeography, 42*(7), pp.1281–1292.

Wood, J.W. and DeWitte, S., 2003. Was the Black Death yersinial plague? *The Lancet Infectious Diseases, 3*(6), pp.327–328.

Wood, J.W., Ferrell, R.J. and Dewitte-Avina, S.N., 2003. The temporal dynamics of the fourteenth-century Black Death: new evidence from English ecclesiastical records. *Human Biology, 75*(4), pp.427–448.

Yeloff, D. and Van Geel, B., 2007. Abandonment of farmland and vegetation succession following the Eurasian plague pandemic of AD 1347–52. *Journal of Biogeography, 34*(4), pp.575–582.

Yoder, C.J., 2006. *The Late Medieval agrarian crisis and Black Death plague epidemic in Medieval Denmark: a paleopathological and paleodietary perspective*. PhD Thesis, Texas A&M University.

Yue, R.P., Lee, H.F. and Wu, C.Y., 2016. Navigable rivers facilitated the spread and recurrence of plague in pre-industrial Europe. *Scientific Reports, 6: 34867.*

Zabalo Zabalegui, F.J., 1968. Algunos datos sobre la regression demografica causad por la peste en al Navarra del siglo XIV. In *Miscelanea Jose M. Lacarra* (pp. 81–87), ed. J. M. Lacarra. Zaragoza: Universidad de Zaragoza Press.

Zhou, D., Han, Y., Song, Y., Huang, P. and Yang, R., 2004. Comparative and evolutionary genomics of Yersinia pestis. *Microbes and Infection, 6*(13), pp.1226–1234.

Ziegler, P., 1969. *The Black Death.* New York: Harper Collins.

Zimbler, D.L., Schroeder, J.A., Eddy, J.L. and Lathem, W.W., 2015. Early emergence of Yersinia pestis as a severe respiratory pathogen. *Nature communications, 6*, p.7487.

5   The scourge of *Y. pestis* re-emerges and persists from 1361 to 1879

Second Plague Pandemic continues

The plagues of Europe between the years of 1361 and 1815 are blamed for the deaths of nearly one third of Europe's population (Alfani 2013). During this 500-year period, plague reappeared every 20–30 years across Europe but in a spatially non-contiguous manner (Schmid *et al.* 2015). In other words, it would reappear at different places and at different times across Europe. In an amazing piece of detective work involving the analysis of 7,711 geo-referenced historical plague outbreaks identified between 1346 and 1859 in Europe and 15 tree-ring records from Europe and Asia, Schmid *et al.* (2015) were able to identify 16 plague outbreaks (1346, 1408, 1409, 1689, 1693, 1719, 1730, 1737, 1757, 1760, 1762, 1780, 1783, 1828, 1830, 1837) that appeared to be reintroductions of plague to Europe from Asia through ports in southern Europe and the eastern Mediterranean. Nine of these reintroductions occurred roughly 15 years after climatic anomalies (warm, wet springs and summers) in Central Asia in 1331, 1394, 1674, 1703, 1723, 1741, 1747, 1766, and 1814 (Schmid *et al.* 2015). The other ~7,600 plague outbreaks appear to have been caused by plague moving from town to town, city to city, port to port within Europe. However, given the lack of rodent reservoirs in Europe, these outbreaks suggest human to human transmission. These data also suggest that the overland Silk Road played a more minor role in plague reintroduction than generally thought. There might be several reasons for this: the Golden Horde or Mongols effectively shut down or severely reduced the overland Silk Road(s) trade traffic in and around 1500, although China did compensate by expanding the maritime Silk Network of shipping lanes (Rezakhani 2010) that bypassed the Asian steppes, and/or plague evolved into multiple strains among four lineages cited as responsible for the cyclical nature of plagues within Europe between 1346 and 1815 (Bos *et al.* 2016; Cui *et al.* 2013; Gage and Kosoy 2005; Haensch *et al.* 2010; Orent 2004; Seifert *et al.* 2016).

Although the sources of plague re-emergence are still debated, its deadly and repeated impact is not in doubt. London, Paris, and Milan were subject to repeated outbreaks of plague. The medieval Black Death returned to London in 1438, 1499, 1511, 1532, 1535, 1563 (with 17,403 deaths in that year), 1578 (with 3,568 deaths), 1592–93 (with 10,675 deaths), 1603 (with 25,045 deaths), 1625 (with

26,350 deaths), 1636 (with 10,400 deaths) (Hall 2008), leading up to the devastating 1664–65 outbreak that resulted in 68,598 deaths in London (Creighton 1894; Hall 2008; Harding 2002). Paris was hit even more frequently: in 1369, 1374, 1380, 1387, 1399, 1412, 1418, 1421, 1432, 1438, 1449, 1466 (with 40,000 deaths; Harding 2002), 1471, 1475, 1481, 1499, 1510, 1519, 1529–33, 1553–55, 1560–61, 1566 (with 25,000 deaths; Harding 2002), 1580 (with 30,000 deaths; Harding 2002), 1595–97, 1604, 1606–08, 1612, 1618–19, 1622–32, 1636, 1638, 1652 and 1668 (Biraben 1975; Harding 2002). Milan was hit more intermittently in 1452, 1468, 1483, 1502, 1523, and 1629–31 (Cohn and Alfani 2007).

Elsewhere, plague erupted in Russia in 1363–65 (in Novgorod and Moscow), 1369–70 (in Lithuania), 1374–77 (in Kiev and Smolensk), 1387–90 (in Pskov, Novgorod, and Smolensk), 1401 (in Smolensk), 1403–04 and 1406–07 (in Pskov), 1408 and 1417–28 (throughout Russia) (Alexander 2002; Melikishvili 2006). The spatial variability of plague in Russia between 1363 and 1428 is probably more due to the irregular and limited chronicling of plague rather than any true random spatial pattern. Thereafter, plague hit Pskov and Novgorod another fourteen times up to the 1552 epidemic (Melikishvili 2006). Fewer plague epidemics erupted in Russia between 1500 and 1700; notable eruptions include 1506 (in Pskov), 1508 (in Novgorod with 15,396 deaths), 1552–53 (in Pskov and Novgorod), 1563 and 1566–68 (in Lithuania), 1570–71 (in Novgorod), 1570–71 (in Moscow), 1592 (in Pskov), 1654 (across central Russia), 1653 (in Crimea and Moscow where half its population died), 1690 (Kursk), and 1692–93 (in Astrakhan) (Melikishvili 2006; Alexander 2002). A plague erupted from the Baltic to Kiev in 1710 during a time of military action, but otherwise in the 18th century, in part due to medical vigilance, localized quarantine, and the use of sanitary cordons, plague was increasingly restricted to eastern Russia (Alexander 2002). Thereafter, plague outbreaks occurred in 1727–28 (in Astrakhan) and in 1738–39 (across Ukraine and southern and central Russia) (Alexander 2002). The widespread plague outbreak of 1738–39 appears to coincide with the Russo-Turkish War, suggesting the social-economic devastation caused by the war provided the setting for the plague to invade Russia. Thereafter, the last great plague hit Russia in 1770–72 and killed hundreds of thousands (Alexander 2002).

This difference in plague outbreak frequency across Europe might be the product of several factors, one being geography, and another the power and efficacy of communities in establishing successful quarantines. Paris, for example, is part of mainland Europe, whereas Britain is more isolated as an island. Navigable rivers certainly facilitated the spread and recurrence of plague in pre-industrial Europe. As a result, those hamlets and towns that were not associated with a navigable river suffered fewer plague outbreaks (Yue *et al.* 2016). Well-maintained city walls were sometimes effective barriers to plague in small towns and cities, but walls built around large cities, ports, and capitals (e.g., Paris, Vienna) highly dependent on trade were generally infective against plague. However, in Edinburgh in October 1574, effective town council regulations that involved closing city gates between 6 p.m. and 6 a.m., guarding them with a watch of six men and isolation of anybody sick within the city, prevented plague in nearby Kirkcaldy and

Leith from entering Edinburgh (Marwick 1574). The Balearic Islands established early and effective quarantine as islands had a natural geographic advantage—isolation. Milan had a powerful and effective government that supported isolation procedures and plague hospitals. But plague was able to circumvent even strict quarantine measures. In 1720, the merchant ship Grand-Saint-Antoine docked in Marseilles. It was immediately quarantined by the port authority, but not before the captain had instigated a voluntary quarantine prior to docking. Nevertheless, a dock guard, assigned to prevent people breaking the quarantine, came down with plague. Over 100,000 died in Marseille and Provence. The port responded by building a Lazaret (a quarantine station) to isolate people and cargo from suspected plague vessels.

Certainly quarantine, first practiced in Pistoia, Italy, though not very effectively initially, (Chiappelli 1887) and the Balearic Islands in 1348 (Arandez 1980) became increasingly effective. In 1630, Pistoia lost just 1.5% or 120 of its 8,000 residents to plague (Cipolla 1981). Large, mainland European cities struggled throughout the 500 years from 1361–1815 to effectively isolate themselves. However, Milan and Venice were successful in implementing quarantine by 1350 (Biraben 1975). Why *Yersinia pestis* ceased to affect Europeans after 1815 remains a mystery, actually; the last place in western Europe to suffer plague was Bari, Italy in 1815 (Morea 1817; Cohn 2008). However, plague continued to rage across Russia and the Middle East until 1879.

The weather, persistent warfare, high taxation, and famines did not help the poorer members of society get through these plagues. Ice-rafting along the North Icelandic Shelf (Eiríksson *et al.* 2006) suggests northern Europe continued to experience cold weather in the form of the Little Ice Age from 1300–1900. The first half of the Hundred Years' War fought between England and France between 1337 and 1453 coincided with the plagues of 1347–1351, 1361, 1374, and 1383. The Battle of Crécy was fought on August 26, 1346, just two years before the plague crossed the channel into England. The Hundred Years' War, subsequent tax increases to pay for the fighting, and associated inflation meant that the 1340s to the 1380s in north-west Europe were a period of tremendous social, economic, and food stress among the poorest, weakest, and most vulnerable members of society. These conditions sadly created a perfect storm for *Y. pestis* to kill large numbers of Europeans.

Changing plague mortality

The plague of 1361 was as lethal as the 1347–1353 primary wave of the medieval Black Death (see Chapter 4). In Avignon in 1348, two-thirds of the population contracted plague, according to the Pope's personal doctor, Raymundus Chalmelli de Vinario, with half the population dying (Cohn 2008). De Vinario also noted that in 1361 half the population caught plague but few lived, while in 1371 only 10% died, and in 1382 only 5% became sick, and most of these individuals lived (Cohn 2008). However, ecclesial records of the Bishop of Winchester in England suggest clergy mortalities dropped from 58.5% in 1349–1350 to 30% in 1361–62 (Mullan

2007). In 1374, in Pisa, the plague killed mostly those aged less than 12; thereafter, chroniclers referred to the plagues of 1361, 1374, and 1383 as the 'plagues of children'. For instance, in Siena, children aged 12 or younger represented 88% of the fatalities in 1383 (Cohn 2008). It would appear that those who contracted and survived the 1347–1351 primary wave of the medieval Black Death obtained immunity from subsequent outbreaks and hence survived in large numbers, while children lacked this acquired immunity. However, in England the mortality was only slightly higher among 6–10 year olds in 1361 (Scott and Duncan 2001). In the Low Countries, the highly localized plague epidemics of 1400–1401 and 1438–1439 killed between 20 and 25% of the population; thereafter, from 1440–1475, plague killed very few people (Blockmans 1980).

Records of mortality in the primary wave of the medieval Black Death indicate large cities, ports, and capitals struggled with the plague longer than smaller cities and towns (Olea and Christakos 2005). As the plague continued to kill between 1361–1815 in Europe, a distinct bias developed. Urban areas suffered higher mortalities than rural areas (Cohn 2008). One suspects that the revival in trade and commerce and increased urbanization compromised human health. In northern Europe between 1500 and 1600, the numbers of people living in urban areas quadrupled (Steckel 2004). Rapid rural to urban migration also intensified wealth, health and diet inequalities (Hoffman *et al.* 2002), and certainly increased urban slum populations. In other words, rapidly expanding urban slums, with their poor, poorly nourished, health-compromised inhabitants living in close, cramped quarters provided optimal conditions for the maintenance and propagation of many zoonoses (animal to human transmitted pathogens), anthroponoses (human to human transmitted pathogens) and sapronoses (water or soil pathogens transmitted to humans). For instance, between 1560 and 1640 plague killed between 8 and 15% of the central European population (Eckert 1996). In Holland during the 17th century, the contrast in urban to rural plague mortalities was not as pronounced because higher rural plague mortalities were associated with conflicts such as the Dutch Revolt and the Thirty Years War, which affected rural areas more than urban areas (Curtis 2016). It is highly likely that these conflicts compromised the diet and health of rural populations as agricultural land was devastated, food stocks and seed grains destroyed or requisitioned by the occupying force, and people forced to move (Curtis 2016).

Although the number of individuals with compromised immune systems expanded as urban populations quadrupled from 1500 to 1600 (Steckel 2004), the return periods of plague lengthened. In fact, plague returned across Europe in the 14th century to locales every 10 years or so, but by the 17th century major cities were suffering from plague roughly once every 120 years (Cohn 2008). As the return period lengthened, so too did the spatial incidence of the plague change – in London in 1563 both rich and poor suffered similar mortalities, but by the 1660s, the poor, working-class peripheral parishes were devastated by plague, while the richer parishes suffered considerably fewer fatalities (Hall 2008). At the same time, between 1520 and the early 1600s, real wages in England declined by more than 40% (Hopkins 1956) and consequently, the poorer parishes got poorer.

Between 1560–61 and 1624–65 the number of vagrant cases at Bridewell's court, East London, rose from 69 to 815, a twelve-fold increase (the majority of these vagrants self-identified as Londoners), although London only grew four-fold during this time period (Beier 1978). In 1598, the Privy Council of London considered parts of Southwark, Shoreditch, Cripplegate, Clerkenwall, Whitechapel, and Houndsditch the "harbors of thieves, rogues and vagabonds" (Beier 1978). I suspect the higher density, crowded housing in the poorer London parishes contributed to the rapid human to human transmission of the plague. In other words, poorer people would be subject to more frequent human interactions, whereas in the lower density, richer neighborhoods, less frequent commingling of people reduced the likelihood of plague transmission. I also suspect that poverty forced many people to carry on working until nearly incapacitated by the plague; in other words, infected people would continue working and infecting others up until death. I also suspect that quarantine laws were not so rigorously enforced in the poorer neighborhoods, in part because of fear of contamination, violence, and the unknown.

The following five locations illustrate the vagaries of plague and the human response to plague in Europe between 1361 and 1815.

The plague and mortality in Penrith (1597–98)

In contrast to Givy's recorded mortality in 1348, which peaked on the 79th epidemic day and finished on the 149th epidemic day (see Figure 4.1), Penrith's epidemic began on September 22, 1597 with the death of a stranger, Andrew Hogson, peaking on the 333rd epidemic day, and finishing on the 457th epidemic day (Scott and Duncan 2001). Of the 1,350 people living in Penrith prior to the plague, 485 died from among 242 families, of which 63 families were completely extinguished, while 79 families lost all but one parent (Scott and Duncan 2001). This occurred even though Penrith was quarantined. The length of Penrith's epidemic is misleading, though; the epidemic suffered two mortality peaks, one in November 1597 followed by no deaths in January 1598 and only one in February 1598, and a secondary, larger peak in June–July of 1598 (Scott and Duncan 2001). The Penrith epidemic therefore appears to mimic Givy's epidemic with a peak in late summer. However, in Penrith's case, the initial epidemic death occurs late in the summer the previous year, plague is dormant through winter with few cases, and then it re-emerges in the spring (Scott and Duncan 2001). This pattern suggests that plague at this time was still a human to human transmitted disease requiring significant human interaction to facilitate transmission (Welford and Bossak 2009).

The plague and mortality in Parma (1630)

It appears that German mercenary soldiers involved in the Thirty Years War introduced the plague into northern Italy in early 1629 (Manfredini *et al.* 2002). The epidemic in Parma began in April, yet quarantine was only enacted on June 19, too late for the epidemic as it peaked in July 1630, while the last plague death was in December of 1630 (Manfredini *et al.* 2002). Excess plague-induced mortality, above the background rate of mortality, suggests that 5–14 year olds suffered ~33% excess mortality, 15–24 year olds ~20% excess mortality, and 25–54 year olds ~11–18% excess mortality (Manfredini *et al.* 2002). These data suggest plague mortality was highest among 5–24 year olds, something seen in London in 1348 during the primary wave of the mBD. The birth rate also dramatically declined by some 39% in the post-plague environment and did not return to pre-plague conditions until 1633 (Manfredini *et al.* 2002).

The plague and mortality in Venice (1630)

During the plague that hit Venice in 1630–1632, women, particularly those who were pregnant, made up 59% of the deaths (Ell 1989). In a 3-day period, 1,163 deaths were recorded between October 23–25: of these, three quarters were plague-related deaths and, of these, 29 were pregnant women (Ell 1989). Although all 77 parishes in Venice suffered plague fatalities, 70% of the deaths occurred in just 20 parishes (Ell 1989). The spatial pattern of plague eruption in Venice suggests that rather than being random in its devastation (which we would expect if driven by rat fleas), plague erupted sequentially among neighborhoods (Ell 1989), just as human to human transmitted pathogens do today (e.g., flu, Vazquez-Prokopec *et al.* 2013; Yang *et al.* 2009).

The plague and mortality in Eyam (1665-66)

Traditionally, a journeyman, Alexander Hadfield, is said to have brought plague to Eyam in a flea-covered bolt of cloth in the late summer of 1665, with George Vicars subsequently dying of plague on September 7 and the second victim Edward Cooper being buried on September 22 (Scott and Duncan 2001). Another explanation is that plague arrived August 20 when people visited Eyam to participate in the Eyam Wakes (Batho 1964; Scott and Duncan 2001). The epidemic killed 260 people, exhibited an initial mortality peak in October 1665, lay dormant through winter, and peaked again in August 1666 (Scott and Duncan 2001). In a repeat of Penrith, Eyam was quarantined and yet suffered many deaths. The epidemic seems to have struck families in a sequential manner, with one family member after another dying before a separate family exhibited the disease (Scott and Duncan 2001; Whittles and Didelot 2016). This series of sequential family infections, again, suggests a human-to-human spread of pneumonic plague, as caregivers and immediate family members (such as seen in Manchuria in the 1910s; Kool and Weinstein 2005) are infected, rather than the spread of rat-to-flea bubonic plague, which would not exhibit a wave-form infection that targeted one family after another. Recent stochastic modeling and Bayesian analytical analysis of Eyam parish data suggest a more complex situation with rat-to-human bubonic plague accounting for 25% of the deaths and human-to-human pneumonic plague accounting for the remaining deaths (Whittles and Didelot 2016). Risk of contracting and dying of plague was also highest among the poor (Whittles and Didelot 2016).

The plague and mortality in Provence (1720–22)

The plague arrived in Marseilles aboard the ship the Grand Saint-Antoine that had previously docked in Saida, Syria on January 30; in Tripoli, Libya, where several Turkish passengers embarked on April 3; and Leghorn, Italy, where port entry was refused on May 17 (Scott and Duncan 2001; Drancourt and Raoult 2002). On May 25, when the Grand Saint-Antoine arrived in Marseilles, port inspectors quarantined both passengers and merchandise on June 3 as a precaution following a voluntary quarantine initiated by the captain because four plague deaths had occurred on board (Scott and Duncan 2001). Although quarantined, the Grand Saint-Antoine discharged its cargo on June 12. Thereafter, on June 20, the first resident of Marseilles fell ill, and the first death (a tailor) occurred on June 28 (Devaux 2013). In all, some 119,811 people died of plague, or 30% of the inhabitants of Provence (Biraben 1975; Signoli et al. 2002). Under normal circumstances, the Grand Saint-Antoine would not have discharged its cargo until the end of a quarantine period. Instead, the ship's owner, the deputy mayor of Marseilles, intervened and had the cargo unloaded (Devaux 2013). Women suffered excess mortality; however, the young aged 0–9 and the elderly over 70, suffered lower mortalities than normal (Signoli et al. 2002). Typically, more than 66% of all plague deaths in communities occurred within the first two months after the first plague deaths were reported, and 61% of all victims were linked to another victim through family ties (Signoli et al. 2002). These data, again, illustrate the importance of family-driven infections that support human-to-human transmission.

The plague and mortality in Russia (1770–1772)

Sadly, the legislation/decrees issued in response to the 1727–28 Astrakhan plague to combat further plague epidemics were not implemented immediately, and in 1770 General Von Shtoffeln (commander of the Moldovan troops that captured the Romanian city of Iasi in January 1770 during the 1769–1774 Russo-Turkish war) failed to enact quarantine and a plague sanitary cordon after his troops developed plague (Melikishvili 2006). Although over 1,500 patients suspected of having plague were isolated in a military hospital, the plague escaped and erupted across Moldavia, Romania, Poland, and Ukraine (Melikishvili 2006). The first plague case in Moscow was documented in November 1770 (Alexander 2002; Melikishvili 2006). Once it reached Moscow, upwards of 100,000 died of plague (Alexander 2002; Cromley 2010; Melikishvili 2006). Here again, bureaucratic indecision greatly aided the spread of plague throughout Moscow. First, Catherine the Great initially refused to accept that plague was advancing on Moscow, and second, although a Russian doctor correctly diagnosed plague among Moscow hospital patients, the German doctor in charge of Moscow's public health system ignored this first report and subsequent reported plague cases (Melikishvili

2006). In fact, it was not until March 10, 1771 that the authorities finally acknowledged the presence of plague in Moscow (Melikishvili 2006), but by that time it was too late. Contempt for the authorities was so widespread in Moscow that on September 5, 1771 Moscow suffered a 3-day plague riot (Melikishvili 2006).

These plague epidemics illustrate several things: one, the human-to-human nature of plague transmission as family members typically died in a sequence; two, the summer-fall seasonal pattern of contagion was typical of the medieval Black Death; three, the risk of contracting and dying of plague was highest among those in poverty; and four, the impact of human error or biased (if not prejudicial) decision-making allowed the plague to erupt and kill hundreds of thousands in Marseilles and Moscow. Sadly, the deadly plague epidemics that ravaged Marseilles in 1720–22 and Moscow in 1770–72 occurred as plague in Europe ceased to be a pandemic threat, as local, regional, national and international public health efforts in the form of quarantine, sanitary cordons, border controls, and legal (and illegal) disease surveillance programs had almost eradicated plague in Europe.

Epidemiology, re-emergence and extinction of the plague between 1361 and 1879

Bubonic plague, historically accepted as the cause for the epidemic outbreaks of medieval Europe, is spread via a rat flea as a vector. However, recent analysis of historical records suggests that a rat-flea-human method of transmission is unlikely for the plagues of medieval Europe (Cohn 2002, 2008; Duncan *et al.* 2005; Christakos *et al.* 2007; Bossak and Welford 2009, 2010, 2015). To recap discussions in Chapters 2, 3 and 4, rats must die en masse to precipitate flea migration to another blood source, in this case humans, to initiate a bubonic pandemic. But a lack of suitable animal reservoirs, i.e., rat, marmot or gerbil, in Europe between 1347 and 1815 (Amori and Cristaldi 1999; Davis 1986; Rielly 2010; Schmid *et al.* 2015) necessitated a re-evaluation of the epidemiology of plague and the medieval Black Death. Today DNA obtained from teeth and bone from skeletons from known medieval Black Death burials across Europe between 1347 and 1815 (Bos *et al.* 2011, 2016; Drancourt *et al.* 2004; Haensch *et al.* 2010; Schuenemann *et al.* 2011; Seifert *et al.* 2016), supports the identification of a plague *biovar, Medievalis*, as the source of the medieval Black Death (Drancourt *et al.* 2004). The plague *Medievalis biovar* (one of several sub-species of plague) carries the protease Pla coding pPCP1 plasmid that allows *Y. pestis* to induce severe pneumonia but not an invasive infection like bubonic plague (Zimbler *et al.* 2015).

In addition to the lack of rats, five additional lines of historic evidence support the notion that the medieval Black Death from 1346 to 1879 was pneumonic. First, the late spring, summer and early fall seasonality of plague infection and mortality mentioned in Chapter 4 persisted through the subsequent European plagues and mirrors the outdoor commingling of people at trade fairs, harvest festivals, and religious pilgrimages (Cohn and Alfani 2007; Welford and Bossak 2009). This is in contrast to modern plague epidemics (e.g., in India) that are restricted to cold weather months when people spend more time indoors (Christakos *et al.* 2007).

Second, descriptions of coughing and splitting blood were regularly noted among doctors, and deaths continued to cluster among families and households (Cohn and Alfani 2007; Scott and Duncan 2001; Signoli *et al.* 2002). Third, during the Second Plague Pandemic, plague and famine went hand-in-hand, whereas, during the Third Pandemic, particularly in India and China, plague coincided with a bumper harvest crop (Christakos *et al.* 2007; Cohn 2008). In these cases, famines compromised health during the medieval Black Death, but during the Third Pandemic, which was mostly a bubonic plague, a bumper crop would increase the number of flea-harboring, plague-infested mammals such as rats. Fourth, the rapidity of plague transmission across England, Europe and Asia also offers further indirect support for human-to-human transmission of the medieval Black Death (Pruitt 2014; Wood and DeWitte-Aviña 2003). During the primary wave of the medieval Black Death, plague transmission velocities varied from 0.9 to 6 km per day (Benedictow 2004; Bossak and Welford 2015; Christakos *et al.* 2005; Noble 1974). During the 1770–71 plague that hit Russia, plague transmission velocities varied between 1.8 and 4.3 miles/day (Alexander 2002). These are far in excess of bubonic plague transmission velocities observed today. In the western USA, plague spreads out from San Francisco and Los Angeles ports at approximately 80.81 ± 15.28 km/year (Adjemian *et al.* 2007), while between 1777 and 1964, plague in China typically spread at less than 40 km/year (Xu *et al.* 2014). Fifth, according to Pruitt's interview with Dr. Tim Brooks, a pneumonic plague researcher at Porton Down, England, Dr. Brooks stated quite emphatically that "It [bubonic plague] cannot spread fast enough from one household to the next to cause the huge number of cases that we saw during the Black Death epidemics" (Pruitt 2014).

Source of plague re-emergence

In Kazakhstan, plague persists and spreads across the landscape once gerbils (one of the reservoirs of plague) reach a gerbil-density threshold, and occupied gerbil burrows form local clusters (Wilschut *et al.* 2015). The transmission velocity and persistence of *Y. pestis* among gerbils is determined both by the distance between gerbil burrows and the distances between clusters of burrows, but it is the clusters that act as foci or metapopulation structures that facilitate a plague outbreak (Wilschut *et al.* 2015). However, at a regional scale, mountain chains in the landscape cause plague to spread in a north-west to south-east direction (Wilschut *et al.* 2013). *Y. pestis* prevalence among gerbils increases during warmer springs and wetter summers than average, with a 1°C increase in spring temperatures leading to a 50% increase in gerbils exhibiting *Y. pestis* (Stenseth *et al.* 2006). In other words, warmer, wetter conditions in spring and summer increase gerbil forage resources, leading to gerbil population growth. At some point, gerbils exceed the gerbil-density threshold (Wilschut *et al.* 2015), allowing *Y. pestis* to reach pandemic portions within the gerbil populations within central Asia, and thus increasing the likelihood of plague transmission to humans. Tree-ring climate proxy data suggest conditions in central Asia were suitable for *Y. pestis* to reach pandemic

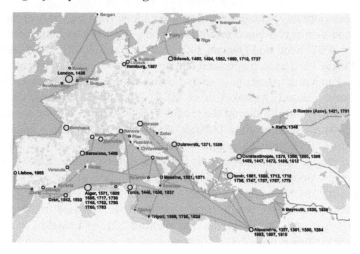

Figure 5.1 Plague outbreaks in maritime harbors of Europe not related to nearby
land-based or maritime harbor outbreaks. (Reprinted with permission from
PNAS from Schmid *et al.* 2015.)

portions within the gerbil populations between 1280 and 1355 and again between
1855 and 1870 (Stenseth *et al.* 2006). This hypothesis supports the occurrence of
the medieval Black Death (Chapter 4) and the origin of the plague in India and
China in the late 1800s (Chapter 6).

Simply put, plague (*Y. pestis*) has its reservoir in central Asia. Warm springs
and wet summers increase small mammal forage, small mammal populations
increase above a small mammal-density threshold, and plague jumps species
to humans. Plague then radiates outward along land and sea trade routes of the
ancient Silk trade route linking central Asia and China, with northern India and
Europe seeding the primary wave of the medieval Black Death from 1347 to 1353
and the emergence of plague in India and China in the late 1800s. However, central
Asian climate scenarios do not seem to be conducive to continuous re-seeding
of repeated outbreaks of *Y. pestis* across Europe between 1361 and 1815 in the
post-primary waves of the medieval Black Death, otherwise known as the Second
Plague Pandemic.

More recent fine-scale resolution of tree-ring climate proxies and analysis of
7,711 historical plague outbreaks across Europe and Asia between 1361 and 1801
by Schmid *et al.* (2015) suggest central Asia reservoirs of *Y. pestis* still seeded
European outbreaks during this 450-year period but on a 15-year delay. Schmid
et al. (2015) also found no evidence for permanent plague reservoirs in Europe.
Of the 7,711 plague outbreaks between 1361 and 1801, some 61 maritime introductions of plague (where plagues had not occurred for two years prior to an
introduction and not within 500 km of a major city or within 1,000 km of another
harbor) occurred in 17 of the 46 principal European trade harbors in the 60 years
between 1346 and 1859 (Figure 5.1; Schmid *et al.* 2015).

In other words, the vast majority of plague outbreaks were caused by transmission among other outbreaks already raging in Europe. People, be they traders, sailors, travelers, or soldiers, moving to and from fairs, festivals, towns and villages, pilgrimage sites, and battlefields (Welford and Bossak 2009; Bossak and Welford 2015) were acting as a pool of infected hosts spreading the plague far and wide. In fact, the latest genomic work suggests there were four *Yersinia pestis* lineages circulating in Europe, part of a big-bang radiation of *Yersinia pestis* (see Figure 3.3; Cui *et al.* 2013), which facilitated repeated plague epidemics from 1346 to 1815 in Europe (see Chapter 4; Bos *et al.* 2016). The diversity of lineages and associated strains of plague was probably the result of selective pressures operating on the plague genome. One possible scenario is that each epidemic (within the Second Pandemic) would kill but also leave survivors whose immune systems would confer immunity to that strain or even lineage of plague. Over time, the number of individuals with acquired immunity would grow and confer greater herd immunity on a population to a particular plague strain/lineage. In other words, over time, a plague strain or lineage would run out of susceptible individuals; only those pathogens that acquired slightly different genomes through the accumulation and selection of beneficial single nucleotide polymorphisms (SNPs) would survive and propagate. Several of these European lineages are unique to Eurasia, but as no non-human reservoir has been identified (Cui *et al.* 2013; Schmid *et al.* 2015), these strains or lineage branches (see Figure 3.3) must have circulated among humans as a human-to-human disease, in other words, as a pneumonic plague. Plague diversity documented among known plague victims within Europe does not rule out regular westerly pulses of plague emerging from Asia (Figure 5.2; Schmid *et al.* 2015) and contributing to plague diversity and subsequent European plague epidemics.

Only 16 outbreaks could not be associated with other plague outbreaks in Europe and, of these, only nine could be associated with major droughts in central Asia following warmer springs and wet summers within a 15-year interval: these were 1331, 1394, 1674, 1703, 1723, 1741, 1747, 1766, and 1814 (Schmid *et al.*

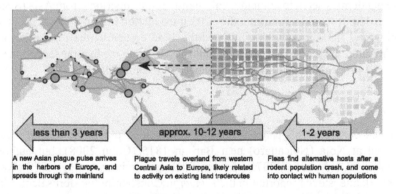

less than 3 years | approx. 10-12 years | 1-2 years

A new Asian plague pulse arrives in the harbors of Europe, and spreads through the mainland | Plague travels overland from western Central Asia to Europe, likely related to activity on existing land traderoutes | Fleas find alternative hosts after a rodent population crash, and come into contact with human populations

Figure 5.2 Map of the plague introduction into Europe. (Reprinted with permission from PNAS from Schmid *et al.* 2015.)

2015). Although Schmid *et al.* (2015) described plague dynamics among gerbils, recent evidence points to marmots being the source reservoir for the Second Plague Pandemic (Bos *et al.* 2016). The 15-year delay between warm, wet springs and summers in Central Asia and the emergence of plague in Europe is the result of plague negotiating the ~4000 km from central Asia to the Black Sea along the myriad of Ancient Silk Route trails operated by camel caravans. Additional delays occurred as the plague pandemic moved from eastern Mediterranean harbors (e.g., Constantinople, Izmir, Alexandria, Alger, Tunis, Oran, Messina, Barcelona, Dubrovnik) to other harbors and into the hinterland of Europe (see Figure 5.2; Schmid *et al.* 2015). In other words, conditions in 1331 in Central Asia led to the 1346 primary wave of the medieval Black Death, while the 1394 Central Asian climatic event possibly led to two reintroductions in 1408 and 1409, and the 1814 event possibly caused reintroductions in 1828 and 1830 (Schmid *et al.* 2015). In all, nine Central Asian climatic events more than likely caused 11 reintroductions of marmot-sourced plague into Europe between 1346 and 1859.

It appears that periodic climate change in the steppes of Central Asia precipitated population explosions among *Yersinia pestis* mammal reservoirs (Schmid *et al.* 2015). Today, marmots, pikas, susliks, and great gerbils are known to act as mammalian plague reservoirs in the central steppes of Kazakhstan (Gage and Kosoy 2005; Schmid *et al.* 2015; Wilschut *et al.* 2015). These population explosions seed *Yersinia pestis* cross-species transmission into humans and then human-to-human transmission precipitates a plague across Central Asia and Europe. Outside of Central Asia, human-to-human transmission occurred because permanent reservoirs of *Yersinia pestis* have not been located (Schmid *et al.* 2015). Indirect support for human-to-human transmission comes from Iceland, which was subject to *Yersinia pestis* plagues, yet rats were absent from Iceland until very recent times (Karlsson 1996).

Even though Europeans developed better plague surveillance, legislation, and enforcement of quarantine, and plague hospitals over time, massive deadly pandemics still raged through Europe in 1575–77, 1629–30, 1656–67, 1709–12, 1719–20, 1743, and 1770–71, killing rich and poor alike (Cohn 2008). Several of these pandemics seem to coincide with plague reintroductions from Asia (Schmid *et al.* 2015). It would appear that the 1703 introduction initiated the 1719–20 epidemic, the 1723 reintroduction initiated the 1743 epidemic, the 1747 reintroduction initiated the 1770–71 epidemic, and the 1814 reintroduction seeded the last western European epidemics of 1828 and 1830 (Schmid *et al.* 2015).

Plague extinction

As noted earlier in this chapter, the last outbreak of plague in western Europe occurred at Noja (Noicattaro) near Bari in 1815 (Cohn 2008), but outbreaks continued in Odessa in 1837 and Astrakhan in Russia as late as 1879 (Walløe 2008). Plague is even reported in the *Lancet* in Constantinople, Jerusalem, Malta, Persia and South China in 1882 (Walløe 2008). The disappearance of plague from Europe demands that three questions be answered: *one*, did the European plague strains and/or lineages go extinct; *two*, did human actions, such as the increasingly effective use of quarantine practices, banish plague from Europe; or *three*, did plague either evolve into, or was replaced by, a less virulent, less deadly, less infectious pathogen? Several historians have questioned whether the plague that hit Rome, Givy, Penrith, Eyam, Marseilles and Astrakhan or that hit India and China and continues to today is one and the same (Cohn 2002; Walløe 2008). Genetically, plague has changed during this period; it has continued to evolve, losing and gaining genetic code along the way (Bos *et al.* 2011; Drancourt *et al.* 2004; Haensch *et al.* 2010; Schuenemann *et al.* 2011; Seifert *et al.* 2016). High-throughput DNA sequencing following array-based enrichment of *Y. pestis* DNA samples from 1346–1815 indicate that the four strains/branches of *Y. pestis* responsible for the medieval Black Death have become extinct (Bos *et al.* 2016). Furthermore, the strain responsible for the Third Pandemic is distinct from the original strain that seeded the medieval Black Death and its plague diversity (Bos *et al.* 2016; Cui *et al.* 2013).

What this suggests then is that plague seems to disappear from western Europe in the early 1800s but the question remains what physical or medical or biological factors kept plague out of western Europe but allowed continued epidemics across Eurasia east of the Balkans until 1879. Acquired plague immunity does not appear to explain the extinction of plague in Europe. Acquired immunity to plague is short-lived, as are current vaccinations, and offer no protection from second attacks (Palmer 1998). Short-lived acquired immunity to plague can explain how during the primary wave of the medieval Black Death (1346–1353), plague traveled from south-east Europe to north-west Europe as a wave of infection (and death) with little or no reinfection behind the advancing wave (Bossak and Welford 2015; Cohn 2008). This leaves quarantine and plague surveillance as possible mechanisms that banished plague from Europe.

Quarantine

One of the first quarantine stations was established by the Decree of 1377 in the independent maritime republic of Ragusa (modern-day Dubrovnik) (Cliff *et al.* 2009) that also established the first permanent health office (Frati 1999). A four-week stay in quarantine was decreed (Frati 1999). In 1403, the first lazaretto, a maritime quarantine station and hospital, was established in Venice on the island of Santa Maria di Nazareth (Cliff *et al.* 2009); this was followed by lazarettos on Sardinia and in Genoa (Hickey 2014). Contrary to popular belief, survival in plague hospitals (or in Italy, lazarettos) was quite high at times. During the 1523 plague, Milan's lazaretto had a 66% survival rate (Cohn and Alfani 2007). However, it appears that the Milan health board moved complete households to San Gregorio lazaretto even if only one household member had the plague (Cohn and Alfani 2007). Such actions probably helped Milan to avoid significant citywide mortality.

Vienna in 1555 implemented one of the first travel restrictions to combat plague, whereby all non-residents had to provide evidence they came from non-plague areas (Velimirovic and Velimirovic 1989). This Infection Decree was revised and expanded in 1562 to include some of the following restrictions: several city gates were permanently closed and others manned by police; persons from plague areas were forbidden entry as were wine-pickers and carters; and beggars and vagabonds were evicted from the city (Velimirovic and Velimirovic 1989). During times of plague, Vienna closed schools and moved all hospital patients suffering from plague to the lazaretto (Velimirovic and Velimirovic 1989). Those who had contact with an infected individual were restricted to their homes for 40 days and a white cross was painted on their home's front door (Velimirovic and Velimirovic 1989).

In a rather interesting twist, it appears that travel bans and quarantine did not need to be absolute to still remain effective. In January 1582, Seville's city council announced travel bans and a quarantine in the face of an advancing plague epidemic (Bowers 2007). But rather than be inflexible, the council continually granted individuals quarantine and travel exemptions (Bowers 2007). As a result, trade continued to flourish and civil order was maintained throughout the quarantine period (Bowers 2007).

During the 1664–1665 London plague epidemic, doctors proved useless in the face of plague; however, public health measures and officials probably saved thousands. Among the orders set forth by the Lord Mayor and Alderman of the City of London were the following decrees:

(1) Examiners are to be appointed to ascertain which houses are infected with plague and to give orders for the shutting up of such houses.
(2) Two watchmen are to be appointed to each infected house, one for the day and one for the night, to prevent anyone entering or leaving it.
(3) Searchers and surgeons are to be appointed to inspect the bodies of all who die and to report whether death was due to plague.
(4) The master of every house is required to report to the examiner without delay any case of plague or of unexplained illness in his home.
(5) Burials must take place between sunset and sunrise.
(6) Every infected house is to be marked with a red cross a foot long in the middle of the door and the words 'Lord have mercy upon us' written above it.
(7) Every infected house is to be shut up with its occupants for four weeks. The watchmen are to 'minister necessaries unto them at their own charges' (if they are able) or at the common charge if they are unable.
(8) The streets are to be kept clean and 'no Hogs, Dogs, or Cats, or tame Pigeons, or Conies be suffered to be kept within any part of the city'.'
(9) No beggars are to be allowed in the streets.
(10) It is ordered 'that all Players, Bear-baitings, Games, Singing of Ballads, Buckler-play, or such like causes of Assemblies of people, be utterly prohibited.'

Quoted from Orders Conceived and Published by James Flefter, Printer to the Honourable City of London (Sloan 1973, p. 273).

Decrees 1–4 and 7, that implemented house quarantine, and decree 10, that banned all human commingling, were probably key to limiting the spread of plague once it penetrated into London.

By the middle of the 17th century, plague quarantine officials in Rome in 1656, in Marseilles in 1719, and in Messina in 1743 could identify the ships and cargo that brought plague to their ports and those who first caught and spread the plague beyond the quarantine stations (Cohn 2008). In 1652, a convention was signed between Tuscany and Genoa regulating trade among their ports in an attempt to combat plague (Cipolla 1981). Even in the absence of a scientific understanding of plague vectors, Venetian civic administrators established networks that exchanged information on plague among the Ionian islands. These networks were fairly successful in combating plague between 1600 and 1797 with lazarettos built in 1588 in Corfu and Zante (Konstantinidou *et al.* 2009). In contrast, plague remained endemic to Greece and the Balkans under Ottoman rule, with epidemics in 1718–1720, 1728–1731, 1733–1740, 1756–1765, 1782–1784, 1787–1789, and 1790–1793 (Konstantinidou *et al.* 2009). In the 15th and 16th centuries, the Ottomans used forced migration to secure border areas in the wake of plagues; many of the resulting urban clusters thrived and became loci for subsequent plague epidemics (Nükhet 2015). It was only in 1830 that the Ottomans finally instituted quarantine measures and, of course, these measures were immediately successful in reducing plague incidence (Konstantinidou *et al.* 2009). According to many foreign observers and travelers prior to 1830, the Ottomans were passive, fatalistic, and indifferent toward plague. Others noted that many Ottomans resorted to fleeing ahead of a plague epidemic (Nükhet 2015).

Plague surveillance

Following in the footsteps of the Venetian government, in 1728 the Austrian-Hapsburg government used their army to establish an extended and effective sanitary cordon along their Turkish-dominated Balkan border (Eckert 2000; Rothenberg 1973; Velimirovic and Velimirovic 1989). Units composed of soldiers and civil personnel patrolled routes leading to major border towns, and controlled access at key road junctions and the entrance to major valleys (Eckert 2000; Rothenberg 1973; Velimirovic and Velimirovic 1989). In non-plague years, some 4,000 people patrolled the border; in plague years some 7,000 people were posted along the border, and if plague reached the border, watch numbers increased to 11,000 (Velimirovic and Velimirovic 1989). Personnel were posted so they could see or hear each other, and anyone trespassing was stopped (Velimirovic and Velimirovic 1989). In addition, paid sanitary cordon spies were sent abroad to document and report on disease outbreaks; the Austrian government was still funding these spies in Turkey through the 1930s (Velimirovic and Velimirovic 1989). In contrast, in 1709 East Prussia failed to establish an effective, leak-proof border with Poland, even though the East Prussian government was well aware of the plague (Eckert 2000).

Even in the absence of plague, in 1710 Great Britain finally established a plague policy. This was substantially updated on the advice of Richard Mead, a physician at St. Thomas' hospital in London, in the amended Quarantine Act of 1721 (Hickey 2014). Mead's 'Short Discourse' on the plague established the criteria for the Act: he suggested that "infected air, infected persons, and the transportation of infected goods" propagated plague (Mead 1722). Tsarist Russia also belatedly implemented plague controls after the 1727–1728 Astrakhan epidemic, even though the medieval Black Death had swept through Russia in 1352, again in 1360–1364 and 1387, and between 1400 and 1600 another 21 times (Melikishvili 2006). The first edict quarantined all foreigners during times of plague and established checkpoints along the Russian-Ukraine border (Melikishvili 2006). Additional edicts were established during the disastrous plague of 1770–1772 that killed thousands in Moscow. These included prohibiting public gathering, the shutting-down of all drinking establishments, shops, inns and factories, the quarantine of all infected people, and a ban on all religious activities (Melikishvili 2006). Doctors were encouraged to wear masks, and those that volunteered for quarantine were paid to remain isolated (Melikishvili 2006). Ultimately these proved successful.

It appears that the establishment of increasingly effective quarantine policies and their enforcement reduced plague eruptions across Europe after 1640 (Eckert 2000). Certainly, by the 17th century, plague eruptions in major cities were occurring only once every 120 years (Cohn 2008). Prior to the mid-17th century, local quarantine was enforced at town or city gates; thereafter states developed communication networks that warned of and monitored plague (Konstantinidou *et al.* 2009) and European states increasingly employed troops to monitor the movement of people and goods during plague epidemics (Eckert 2000; Melikishvili 2006).

Although travel bans and quarantines served to reduce the mortality and spread of plague, it appears quarantine was also used by city councils to separate the poor from the wealthy (Carmichael 2014). For example, in Florence after edicts released in 1450, pesthouses and lazarettos were filled with the poor, whereas the wealthy who were possibly infected with plague were confined mostly to their homes (Carmichael 2014).

Early plague historians also argue that the transition to burial in wooden coffins for the plague dead across all classes, rather than in mass graves during plague epidemics, served to extinguish the plague in Britain by the 1700s (Creighton 1894). This argument fails to consider improvements in healthcare and medicine, diet, and sanitation, or the evolution of quarantine practices. The 1831–1854 cholera outbreak that killed my great, great grandfather, Jesse Welford, on August 10, 1849 in Dinton, Buckinghamshire, suggests sanitation was still a huge issue until Snow's 1854 study made the connection between well contamination and human sewage (Snow 1854).

CCR5-Δ32 gene mutation

Identification of a white blood cell gene mutation confined to Europe, termed the CCR5-Δ32 deletion (Martinson *et al.* 1997; Galvani and Slatkin 2003;

Mecsas *et al.* 2004), and current CCR5-Δ32 occurrences that mirror the known intensity and spatial extent of the mBD suggest that the mBD mortality acted as a selection event in forcing modern population representation of this allele mutation (Duncan and Scott 2005). A direct living descendant of an Eyam Black Death epidemic survivor has been found to carry this mutation (Duncan and Scott 2005). It has been hypothesized by Duncan *et al.* (2005) that the medieval Black Death provided a selection pressure through repeated epidemics that selected for the prevalence of the CCR5-Δ32 gene deletion (Duncan *et al.* 2005). However, the CCR5-Δ32 mutation does not confer immunity to *Y. pestis,* which is a bacterium (Mecsas *et al.* 2004). It is also important to note that the CCR5-Δ32 mutation was discovered not during an investigation of mBD causality, but rather as an anecdote that a significant percentage of Europeans were resistant, and in some cases nearly immune, from infection with HIV, *a virus.* Galvani and Slatkin (2003) suggest that repeated outbreaks of smallpox through the medieval period provided the historical selection pressure. The fact that the CCR5gene codes for the surface of white blood cells and offers immunity to smallpox suggests a more plausible hypothesis. In fact, the CCR5-Δ32 gene deletion eliminates the surface receptor CC-CKR-5 used by HIV-1 and smallpox to enter white blood CD4(+) cells resulting in complete resistance to these two viruses (Alkhatib *et al.* 1996; Dragic *et al.* 1996; Galvani and Slatkin 2003). This explanation follows Occam's razor – it is simple and neat – as opposed to trying to explain how *Yersinia pestis,* a bacterium, would cause a gene deletion selection pressure.

Medical knowledge vs. public health

Throughout the medieval Black Death, an interesting dichotomy emerged: the medical understanding of the contagious properties of plague lagged far behind civic authorities' perception of plague infectivity (Cipolla 1981; Palmer 1978). However, one medical innovation did emerge – housing patients with similar ailments in their own ward (Kelly 2006). By 1600, public health officers in Pistoia, Venice, and several northern Italian states were adamant in their belief that plague spread from person to person (Cipolla 1981; Palmer 1978). In contrast, physicians in the 16th century were still supporting classical medical theory (utilizing ideas from Aristotle and Galen) by suggesting the plague was the product of corrupted air. The Counter-Reformation served to re-enforce these dated ideas (Palmer 1978). Even in the late 1700s and early 1800s, French physicians and German officers working in the Ottoman Empire blamed second-hand clothing bought and sold in flea markets run by Jews for harboring a *miasma* that fueled a plague epidemic (Nükhet 2015; Pouqueville 1805).

Nevertheless, the success of localized quarantine established by local city councils paved the way for state-run quarantine programs; in fact, the coordination across broad geographic territories of quarantine edicts, sanitary cordons, and plague-spy networks necessitated that states play a more aggressive and assertive role in public health in Europe and across Russia (Alexander 2002; Van Damme and Van Lerberghe 2000).

The end of plague

Today in Europe, plague is rare, with the last plague epidemic occurring in Bari, Italy in 1815. In other words, plague has become extinct in Europe; the four marmot-derived plague strains/lineages (Bos *et al.* 2016; Figure 5.1; Cui *et al.* 2013; Spyrou *et al.* 2016) that initiated and caused the medieval Black Death from 1346–1815 have disappeared. I believe that the principal reasons for plague extinction in Europe were the aggressive role played by public health officials, the use of quarantine and lazarettos, the use of government-run spy networks that identified and monitored plague locations, and the spatial expansion of quarantine from highly localized entities to broad, border-long sanitary cordons. Elsewhere today, a less infectious plague remains; it is a rat-derived plague (a bubonic plague), one that is blood-borne and hence among humans difficult to catch. Today bubonic plague is sustained by a multitude of animal reservoirs across the planet; plague is neither communicated nor sustained in any large measure by humans.

Recent modeling and laboratory work suggests that long-distance dispersal among diseases (pathogens) selects for higher virulence, while transmissibility declines where there is more local interaction and localized transmissions (Boots and Mealor 2007). This could explain the variation in pandemic lethality and transmissibility observed among the plagues that erupted in Europe from 1346 to 1814. The 1346–1353 plague pandemic was initially highly lethal and transmissible, as it seems were later reintroduced plagues, whereas plague(s) that were self-sustained in Europe appear to be less lethal and less infective. In fact, 11 reintroductions of marmot-sourced plague into Europe and Russia seem to have occurred between 1346 and 1859 (Schmid *et al.* 2015). These include the 1346–1353 pandemic, and both the highly lethal 1720–22 and 1770–72 epidemics (Schmid *et al.* 2015). This hypothesis is very difficult to prove conclusively because the rapid imposition of quarantine, isolation of plague patients in plague hospital wards, sanitary cordons, stricter border controls, and better monitoring reduced many later plague eruptions to short-lived episodes that affected limited spatial areas.

Impact of plague on daily life during the Black Death

For those who survived a plague epidemic, the urban centers offered new more lucrative employment, the opportunity to move into and up within trading and manufacturing guilds, and the chance to move into new, abandoned houses and businesses. As a result, upswings in rural-urban migration characterized medieval life after each plague epidemic (Cohn 2007). Several countries and communities passed legislation to defeat this movement, probably fearful of the rise of a new powerful and greedy urban elite (Cohn 2007). Yet the scale of the losses necessitated rural-to-urban migration to fill the voids in the urban populations. For example, Martigues in France, with a pre-plague population of roughly 6,000 inhabitants, lost 2,150 inhabitants between 1720 and 1722, while Aubagne lost 2,114 of nearly 4,000 inhabitants during the same epidemic (Signoli *et al.* 2002).

Research from northern Italy shows that throughout the medieval Black Death, wealth remained concentrated in the hands of an elite located within towns and cities even during and after plague epidemics (EINITE; Alfani 2014). Moreover, wealth inequalities between and within rural and urban areas increased across northern Italy even during periods of regional economic stagnation, and the migration of peasants to urban areas acted to maintain and feed these wealth inequalities (EINITE; Alfani 2014). I suspect that depopulation of rural areas by more upwardly mobile peasants would leave their home communities more impoverished, while these same peasants would enter the lowest rungs of the medieval urban social strata and become members of the unskilled workforce within urban areas. This is a process we see today among many immigrant communities (Boucher *et al.* 2009; Martine and McGranahan 2013). However, many of today's immigrants send home money digitally as remittances, something not available to those migrants during the medieval Black Death.

Demography of Europe during the plague

The medieval Black Death increased the death rate due to plague deaths and led to a decline in the birth rate as plague-initiated socio-economic collapses undermined the European psyche (Scott and Duncan 2001). By 1400, Europe's population was estimated to be half what it had been in 1345 (Scott and Duncan 2001). Surviving medieval church, census, and tax records attest to this; in fact, Europe's population took approximately six generations to recover (Scott and Duncan 2001).

Following the primary wave of the medieval Black Death of 1346–1353, the number of epidemics that hit Europe declined until the early 1400s. Thereafter, the number of outbreaks increased to a peak around the 1660s and then declined precipitously to a consistently low number of outbreaks from the 1680s through to the 1800s (Voigtländer and Voth 2012). In contrast, life expectancies across Europe show an alternating pattern, increasing and decreasing, with the lowest life expectancies coinciding with plague outbreaks of the 1650–1660s and 1730s, yet life expectancy declined from just over 38 to 35 years old from 1551–1746 (Voigtländer and Voth 2012). However, nobles were hit much less by the Black Death (1346–1353) and subsequent plague waves; in fact, their life expectancy increased from 50 to 55 by 1400 (Cummins 2014). This early rise in life expectancy suggests that the nobility in 1400, with their wealth and rich diets, represent the beginning of Europe's demographic transition. This demographic change predates both the industrial revolution and modern medicine and appears to suggest that the medieval Black Death was quite influential in the Rise of the West and the development of the divergence among north-west/south-east European countries (Pamuk 2007; Alfani 2014).

As noted in earlier chapters, the geographic connectedness of towns and cities to trade routes had a significant impact on plague mortality; highly connected towns and cities exhibited higher plague mortalities and longer epidemics. This pattern persisted throughout the plague's presence in Europe up to 1815 and is illustrated by Wellington, Wem, and Whitchurch parishes in Shropshire, England in the 17th century (Watts 2001). All are large parishes between 11,000 and 15,000 acres, all have small market towns, and all are composed of agricultural land, and yet Whitchurch suffered high plague mortalities and its population declined through the 17th century because Whitchurch was located on a major thoroughfare for the trade of cheese and cattle between Chester, North Wales, Ireland, and London (Watts 2001). Its location meant that Whitchurch was repeatedly hit in 1642, 1648, 1649, and 1667 by plague, while the other parishes suffered much lower mortalities, or on occasion avoided the plague altogether (Watts 2001).

Socio-economic impact of plague during the medieval Black Death

Rather surprisingly, major demographic shocks due to plague seemed not to trouble the long-term growth trajectory of Italian cities between 1300 and 1861 (Bosker *et al.* 2008). Yet substantial wage gains after the initial 1347–1351 plague shock seem to be perpetuated by high rural-to-urban migration rates between 1361 and the 1850s across Europe (Voigtländer and Voth 2012). Higher urban wages and more disposable income increased demand for manufactured consumer goods, and, as a result, new industries arose to meet the demand, and urban manufacturing centers grew in size. However, this urban growth increased the number of susceptible plague victims living in densely populated, unhealthy urban centers, thereby increasing urban death rates, especially among the poor, from plague and other opportunistic diseases. This necessitated continuing rural-urban migration to replace dead and dying urban dwellers and fostered and sustained higher incomes. These higher incomes meant two things: more taxable income, which facilitated more borrowing by the rich and politically motivated individuals to sustain and expand trade; and more income to support wars. Both the increased trade and warfare sustained more plague epidemics (Voigtländer and Voth 2012). In other words, it would appear that the rise of Europe is tied to plague shock and the increase in the cost of wage labor, before productivity increases were achieved through technological change and subsequent rapid urbanization (Voigtländer and Voth 2012).

Across Europe the impacts of sustained urban population growth and higher wage labor in urban centers were not equally distributed. So, although major demographic plague-induced shocks to Italian cities do not seem to have troubled their own growth trajectories (Bosker *et al.* 2008), a severe plague-induced demographic crisis in the 17th century in Italy does seem to have been severe enough to cause Italian economic growth to lag behind that of the emerging economies of northern Europe such as the UK, the Netherlands, and Belgium (Alfani 2014; Pamuk 2007). It would appear that extreme plague mortalities observed in both rural (which was not typical during plague epidemics) and urban areas in Italy in the 17th century curtailed urban growth, wage growth, and early industrial innovation and development in Italy (Alfani 2013).

In the immediate post 1346–1351 era, the loss of a significant portion of the agricultural waged labor led to the remaining peasants demanding and getting higher wages as their labor became more valuable (see discussion in Chapter 4). Many peasants in England bought or occupied abandoned rural land to farm themselves (Dyer 2003). This was greatly resented among the rural elite, the landowners, and the early industrialists – the millers, brewers and master weavers. Most landowners, faced with labor shortages and rising wages, tried through manor courts to re-establish social and economic control by the reimposition of serfdom and its tenant service obligations (Kitsikopoulos 2002). Over the next 30 years, the newly liberated serfs and small rural owners resented and tried to resist attempts by the rural elite to re-establish control over their lives, which culminated in England in the Peasant Revolution of 1381 (Kitsikopoulos 2002).

Serfs and real wages

The most immediate effect of the Black Death was a shortage of labor. Much land could no longer be cultivated. In response, the nobles refused to continue the long-established common practice of gradually eliminating serfdom by allowing the serfs to buy their freedom. Over the centuries some people realized that free tenants were more productive than serfs, and this had led to a gradual breakdown in the use of serfs. With the post-Plague labor shortage, many nobles tried to reverse the process in order to keep their land under cultivation and their incomes up. Free tenants took advantage of the labor shortage to demand better terms from their landlords, and nobles were reluctant to see their incomes reduced. Governments tried to fix wages, but the labor shortage was irresistible. If their feudal lords would not relent, serfs simply fled to areas where wages were higher or land rental terms lower. This contributed to significant wage inflation (wages nearly doubled in the immediate aftermath of the 1346–1351 plague; Voigtländer and Voth 2012) and to sustained rural-to-urban migration and growth of cities, while incomes rose by a third between 1500 and 1700 in north-west Europe (Allen 2001). Yet real wages began to stagnate in Spain and Italy somewhere between 1450–1500 (Pamuk 2007), while urbanization rates and growth in output remained well below north-west European rates (Allen 2001). Sometime around 1450, as plague mortality rates started to decline and European populations and outputs (manufacturing and agricultural) increased, wages and per capita incomes began to decline (Pamuk 2007). In fact, the plagues of Europe might have seeded the north-west/south bifurcation of Europe, initiating the division of Europe today, where north-west Europe is wealthier and more urbanized. In north-west Europe, real wages rose higher than south/south-east Europe before 1450 in the post 1346–1353 environment, and from 1450 to 1600 real wages declined more slowly in north-west Europe than south/south-east Europe. As a result, a wage gap between north-west and south/south-east Europe developed (Pamuk 2007). The growing 'wage gap' appears to mirror the increase in urbanization and productivity observed in north-west Europe. North-west Europe's increase in productivity appears to be tied to greater institutional flexibility, with guilds relaxing membership requirements (with more women joining guilds) and apprenticeship rules, the use of labor-saving devices (e.g., the printing press), an expansion of rural industries particularly in Britain and the Low Countries, and an expansion of the female labor force (Epstein 2002; Pamuk 2007).

End of serfdom, the beginning of state border controls and democracy

As noted in Chapter 4, the Black Death sped up changes in medieval society that were already under way. The shock of the plague caused many peasants to demand a restructuring of society, and with it a curbing of aristocratic rights and privileges. When these hopes for a better life were curtly dismissed, or savagely repressed by the nobility through the use of England's 1363 law banning silk, silver buckles, and fur-lined coats, and the 1377 poll tax (Kelly 2006), many commoners rose in rebellion. The French revolts of 1358, 1381, and 1382, the rebellion in Ghent in 1379, and the English Peasants' Rebellion in 1381 were only the tip of the iceberg. Unrest was everywhere. None of the rebellions were successful, but in the end the disintegration of the manor system of managing agriculture began, as the value of labor and capital rose in the aftermath of the Black Death and subsequent waves of plague that hit Europe (Haddock and Kiesling 2002). A land rent system, with the freedom of the peasants recognized, replaced the serf system. In other words, plague's mortality crises seeded a revolutionary shift in land tenure (Haddock and Kiesling 2002). This system still exists in many parts of Europe, although the desire of peasants to own their land eventually led, centuries later, to migration to places such as Russia, Australia, Africa, and the Americas. There was never enough land, and dividing it among the sons soon led to economically unviable situations.

In addition, the imposition of community- and state-supported quarantine, sanitary cordons, stricter border controls, and plague surveillance all contributed, among other factors, to the increased role that states played in the lives of their people. The nature of states evolved with the threat of plague, and states became more intrusive and restrictive in their attempts to control plague (Slack 1988; Watts 2001).

In contrast to the economic and political strife that developed between peasants and aristocracy across much of Europe following the primary wave of the medieval Black Death (see Chapter 4), Holland's economy boomed, with little attendant labor strife (van Bavel et al. 2004). Several factors were at work to create this robust economy. One, although the mBD spread across the Low Countries, fewer Dutch died in the mBD than other comparable populations (Benedictow 2004). Two, much of Holland was occupied just before mBD struck because of the cost and technological challenge of draining and desalinating land reclaimed from the sea (van Bavel et al. 2004). Rather than landed aristocracy acquiring more land, in Holland peasants occupied the new land, established land title, and invested in technologies to drain the land. As a result, the Dutch were quick to adopt other capital-intensive technologies, invest in high-return cash-crops, and pay higher wages (van Bavel et al. 2004). The Dutch also resisted the implementation of a feudal system that would have limited rural-to-urban migration; instead, the Dutch created a very fluid labor market, where trade and city guilds had few powers (van Bavel*et al.* 2004). In fact, in the post-primary wave of the mBD, the towns and cities of Holland were very open to economic opportunity and adoption of new ideas, and new technological innovations. As a result, Holland's economy boomed and urbanization exploded. By the 1570s, 45% of Holland's population lived in urban areas, while 40% of its workforce was in industry, 20% in services, 15% in fishing, and only 25% in agriculture (van Bavel*et al.* 2004).

Gendering of the Black Death

Contrary to popular belief that the Black Death helped begin the liberatation of women, the economic gains acquired in the immediate post-primary wave of the medieval Black Death in the late 15th century were not sustained because women did not acquire more legal rights or political power to protect these economic gains (Goldberg 1988; Mate 1998; Rigby 2000). In fact, rather than occupying more empowering positions within the workforce, women remained in irregular low-skill, low-status, low-paid jobs (Bennett 1992; Rigby 2000). Moreover, the vast class and wealth differences that divided late medieval Europe so ostracized women from each other that no common group identity could emerge to fight for more legal rights and/or political power (Mate 1998).

Taxation, warfare, innovation and manufacturing

Following the primary wave of the mBD and repeated return of mBD, land was abandoned, rents went unpaid, and those who survived demanded higher wages (as mentioned above). The ranks of regular and mercenary soldiers were thinned, but their wage demands rose, yet tax revenues decreased, in some places by more than 50%. As a result of the medieval Black Death, the French converted to a smaller, paid, professional army, something the English had instigated some time before. The primary wave of the medieval Black Death and subsequent plagues also undermined the authority of kingship; the Black Death showed people that kings were human, too, not God's appointed leaders. Both Edward III and Edward the Black Prince fled London for their remote, plague-free estates, while Richard II was forced to abdicate in 1399 (Cantor 2001). Richard II's successors, Henry IV, Henry V, and Henry VI fared little better, even though Henry V was crowned king of France in 1422 (Cantor 2001). It would appear that the increased costs of maintaining large armies and expensive campaigns (especially those abroad) undermined Henry VI's ability to successfully finish the Hundred Years War, with the result being that the English lost vast areas of France (Cantor 2001). In many ways, the failure of the Hundred Years War was a defining moment for the English; it turned them away from trying to maintain a continental European empire, and it also demystified the monarchy. So just as the Justinianic Plague changed the trajectory of Rome, so the aftermath of the primary wave of the medieval Black Death and subsequent plagues quite possibly changed England's imperialist ambitions.

The higher wage environment of the mBD world also offered increased opportunities for inventive and capable individuals, who created labor-saving devices (such as Gutenberg who invented the printing press in 1453) and various people who developed boats that demanded fewer sailors and could sail further (Herlihy 1997). Black-powder and ball portable hand-held cannons, or Schioppo, were first employed in Italy in the later 13th century; they allowed fewer soldiers to kill more men (Herlihy 1997). Demand for luxury goods increased as European societies witnessed increased inequality among its surviving populations (Hoffman *et al.* 2002; Pamuk 2007). In many respects, the medieval Black Death killed off the feudal, restrictive medieval society, with its serfs and protected guilds, and gave birth to the beginnings of our own industrialized consumer society.

Art and language

New themes emerged in art after the Black Death. The arrow, already a symbol of divine punishment, became a symbol of plague infection, and these are typically pictured puncturing the buboes (e.g., the Black Death fresco of 1355, the church of St-André, Lavaudieu, France; DesOrmeaux 2007). In other words, God's punishment was the plague. Possibly for the first time, representations of the manifestations of disease, in this case buboes, were widely illustrated in bibles (e.g., the Toggenburg Bible of 1411), paintings (e.g., Josse Lieferinxe's St. Sebastian Intercedes during the Plague of Pavia, 1497–1499; St. Roch Cured by an Angel), and frescos (e.g., Chapel of St. Sebastian, Lanslevilard, Savoy) (DesOrmeaux 2007). Images of death and destruction came to dominate art.

Figure 5.3 Exterior fresco on the Voroneţ Monastery, Romania. (The church was built in 1488, while the frescoes were painted in 1547 CE; @Mark Welford 2005.)

Foremost among these were paintings created by Hieronymus Bosch, especially The Garden of Earthly Delights (1490–1510), Pieter Bruegal the Elder, especially the Triumph of Death (1562–1563), and the painted monasteries of north-eastern Romania, especially Voronet and Humor with their vivid depictions of death (Figure 5.3). Less well-known images include the Wooden Churches of Maramures, Romania (Figure 5.4), while more mundane representations of death include The Limbourg Brothers' Institution of the Great Litany (DesOrmeaux 2007). Images that appear as *memento mori* include the Bridal Pair from Germany, painted in 1470, depicting a celebration of marriage juxtaposed to an image of the couple dying (DesOrmeaux 2007). Similarly, gravestones and burial monuments were carved with ideal representations of the person usually resting on a skeletal body; these are known as a *transi* or cadaver tombs (DesOrmeaux 2007). Excellent examples include the Tomb of Cardinal Lagrange, Avignon, constructed in 1402 (DesOrmeaux 2007); the tomb of Louis XII and Anne de Bretagne in the Basilica of St. Denis (Figure 5.5); the tomb of John and Katherine Denston, Parish Church of St. Nicholas, Denston, Suffolk that illustrates the partially rotting bodies of this husband and wife; and possibly the earliest *transi* – the tomb of Guillaume de Harcigny at Laon, France from 1392.

Monks, teachers, professors and students of the monasterial schools, colleges, and universities of England died in large numbers in the Black Death. Most of them spoke French, but their replacements spoke mostly English, and so by the 1380s Chaucer was able in the *Canterbury Tales* to ridicule even the elite – the nobility and ecclesial leaders – for their poor French (Cook 1964). Although many contest that the Hundred Years War initiated this revolution in language use in England, certainly the Black Death was partially responsible.

Figure 5.4 The interior of the Saint Parascheva Church, Poinile Izei, Maramures, Romania built in 1700. (@ Andrei Le Guide 2005.)

Figure 5.5   The tomb of Louis XII and Anne de Bretagne in the Basilica of St. Denis.
(Unknown photographer.)

Minorities

Although Jews were suspected of transmitting plague during the primary wave of
the medieval Black Death (see Chapter 4) and subjected to mass violence through
targeted pogroms, rather surprisingly, subsequent plague pandemics from 1361–
1879 did not unleash waves of hatred and violence against Jews or any other
minorities (Cohn 2012). Instead, violence was directed toward a broad spectrum
of people and professions. Nevertheless, health workers (particularly doctors,
and those officials implementing quarantine measures) and those who handled
the dead (the drivers of dead-carts who retrieved the dead each night and the
gravediggers who buried the dead) were targeted in particular and accused of
sustaining these pandemics (Cohn 2012). Scapegoating of people for individual
financial and political gain also occurred; in Geneva, this was first observed in
1545 (Naphy 2002). In a surprising twist, the poor were rarely blamed or victim-
ized, even though the poor and malnourished suffered higher plague mortalities
(Cohn 2012; DeWitte and Wood 2008). This tolerance evaporated in Florence
during the plague of 1575–1578, where the authorities prohibited "tricksters, gyp-
sies, negroes, knaves, herbalists, street-singers, comedians, whores, and similar
oddballs (e simil sorti de genti stravaganti)" from entering the city (Righi 1633;
Cohn 2012, p. 537).

However, during the same plague in Milan, Jews and the poor were not
expelled, beaten or killed; instead officials tried to enforce domestic cleanliness

(Cohn 2012). In 1420, Volterra successfully petitioned Florence to allow Jews to return as their community sought the means to obtain loans to cope with Florence's excessive taxation (Cohn 2007). This illustrates the schism that enveloped much of Europe; commoners frequently supported Jews while the elite sought to expel them (Cohn 2007).

In western Europe, new farming edicts in the mid-14th century from the Pope and Rome, forced Jews off the land and into increasingly segregated ghettos (Cantor 2001). Subsequent plague-inspired pogroms, particularly in Germany, saw Jews move eastwards, as more enlightened monarchs such as Casimir II of Poland encouraged Jewish emigration into Poland (Cantor 2001). As a result, by the early 17th century half the world's Jewish population lived in Poland and western Ukraine under Polish hegemony.

Environmental change

This is discussed at length in Chapter 4.

How was religion affected by the Black Death

Although religious feelings intensified, dissatisfaction with the established (Catholic) church increased because, like other medieval institutions, the church failed to protect its members from the plague. This prompted many of the surviving elite (merchants, artisans and guilds) to invest in private chapels, something Cantor (2001) referred to as the "privatization of Christianity" (Kelly 2006). Certainly, the medieval Black Death was one of several factors (i.e., Henry VIII proclivities) that encouraged religious dissent and open criticism of the Catholic Church across Europe and provided a receptive environment for Martin Luther and his Ninety-Five Theses of 1517.

Summary

Looking back at this time in human history, we can see that the expansion of the Chinese Empire out of Asia indirectly and unconsciously disseminated plague across Europe. Four phenomena between 200 BCE and 1800 CE changed world history, world culture, world trade and world geography: the Silk Road, the Mongol expansion, the expansion of major sea routes connecting China and India and the eastern Middle East, and the medieval Black Death from 1347–1879. At the heart of these phenomena is China – the world's first great trading superpower – who from the 2nd century CE through to the 1700s CE, dominated world trade. Beginning with Emperor Han Wudi and his envoy, Zhang Qian, who instigated diplomatic relations with the Xiongnu of western China, the Silk Road was established and remained a viable trading axis until Mongol expansion necessitated the expansion between 900 and 1500 of sea routes across the China Sea and Indian Ocean (Liu and Shaffer 2007). Coincidently, and rather tragically, the Silk Route – the conduit for trade, technology and people in and out of China – intersected

Central Asia with its pneumonic plague reservoirs. So just as the world achieved its third great transition, climate fluctuations in Central Asia resulted in pneumonic plague eruptions among marmots, which were eventually inflected upon humans.

References

Adjemian, J.Z., Foley, P., Gage, K.L. and Foley, J.E., 2007. Initiation and spread of traveling waves of plague, Yersinia pestis, in the western United States. *The American Journal of Tropical Medicine and Hygiene*, 76(2), pp.365–375.

Alexander, J.T., 2002. *Bubonic plague in early modern Russia: public health and urban disaster*. Oxford, UK: Oxford University Press.

Alfani, G., 2013. Plague in seventeenth-century Europe and the decline of Italy: an epidemiological hypothesis. *European Review of Economic History*, 17(4), pp.408–430.

Alfani, G., 2014. Back to the peasants: new insights into the economic, social, and demographic history of northern Italian rural populations during the Early Modern period. *History Compass*, 12(1), pp.62–71.

Alkhatib, G., Combadiere, C., Broder, C.C., Feng, Y., Kennedy, P.E., Murphy, P.M. and Berger, E.A., 1996. CC CKR5: a RANTES, MIP-1α, MIP-1β receptor as a fusion cofactor for macrophage-tropic HIV-1. *Science*, 272, pp.1955–1958.

Allen, R.C., 2001. The great divergence in European wages and prices from the Middle Ages to the First World War. *Explorations in Economic History*, 38(4), pp.411–447.

Amori, G. and Cristaldi, M., 1999. Rattus norvegicus. In *The atlas of European mammals* (pp. 278–279), ed. A.J. Mitchell-Jones. London: Academic Press.

Arandez, A.S., 1980. La reconquista de las vías marítimas. *Anuario de Estudios Medievales*, 10, p.41.

Batho, G.R., 1964. The plague at Eyam: a tercentenary re-evaluation. *Derbyshire Archaeological Journal*, 84(1), p.81.

Bavel, van, J.P., Zanden, V. and Luiten, J., 2004. The jump-start of the Holland economy during the late-medieval crisis, c. 1350–c. 1500. *The Economic History Review*, 57(3), pp.503–532.

Beier, A.L., 1978. Social problems in Elizabethan London. *The Journal of Interdisciplinary History*, 9(2), pp.203–221.

Benedictow, O.J., 2004. *The Black Death, 1346–1353: the complete history*. Suffolk, UK: Boydell & Brewer.

Bennett, J.M., 1992. Medieval women, modern women: across the great divide. In *Ale, beer and brewsters in England: women's work in a changing world, 1300–1600*, ed. J.M. Bennett. Oxford: Oxford University Press.

Biraben, J.N., 1975. *Les hommes et al peste en France et dans les pays europeens et mediterraneens*. Paris (France): Mouton. Volume 1 (455 p.), Volume 2 (416 p.).

Blockmans, W.P., 1980. The social and economic effects of plague in the low countries: 1349–1500. *Revue belge de philologie et d'histoire*, 58(4), pp.833–863.

Boots, M. and Mealor, M., 2007. Local interactions select for lower pathogen infectivity. *Science*, 315(5816), pp.1284–1286.

Bos, K.I., Herbig, A., Sahl, J., Waglechner, N., Fourment, M., Forrest, S.A., Klunk, J., Schuenemann, V.J., Poinar, D., Kuch, M. and Golding, G.B., 2016. Eighteenth century Yersinia pestis genomes reveal the long-term persistence of an historical plague focus. *Elife*, 5, p.e12994

Bos, K.I., Schuenemann, V.J., Golding, G.B., Burbano, H.A., Waglechner, N., Coombes, B.K., McPhee, J.B., DeWitte, S.N., Meyer, M., Schmedes, S. and Wood, J., 2011. A draft genome of Yersinia pestis from victims of the Black Death. *Nature, 478*(7370), p.506.

Bosker, M., Brakman, S., Garretsen, H., De Jong, H. and Schramm, M., 2008. Ports, plagues and politics: explaining Italian city growth 1300–1861. *European Review of Economic History, 12*(1), pp.97–131.

Bossak, B.H. and Welford, M.R., 2009. Did medieval trade activity and a viral etiology control the spatial extent and seasonal distribution of Black Death mortality? *Medical Hypotheses, 72*(6), pp.749–752.

Bossak, B.H. and Welford, M.R., 2010. Spatio-temporal attributes of pandemic and epidemic diseases. *Geography Compass, 4*(8), pp.1084–1096.

Bossak, B.H. and Welford, M.R., 2015. Spatio-temporal characteristics of the medieval Black Death. In *Spatial analysis in health geography* (pp. 71–84), eds. P. Kanaroglou, E. Demelle and A. Paez. Farnham, UK: Ashgate Press.

Boucher, S., Stark, O. and Taylor, J.E., 2009. A gain with a drain? Evidence from rural Mexico on the new economics of the brain drain. In *Corruption, development and institutional design* (pp. 100–119), eds. J. Kornai, L. Mátyás and G. Roland. Basingstoke, UK: Palgrave Macmillan.

Bowers, K.W., 2007. Balancing individual and communal needs: plague and public health in early modern Seville. *Bulletin of the History of Medicine, 81*(2), pp.335–358.

Cantor, N.F., 2001. *In the wake of the plague: the Black Death and the world it made.* New York: Simon and Schuster.

Carmichael, A.G., 2014. *Plague and the poor in Renaissance Florence.* Cambridge, UK: Cambridge University Press.

Chiappelli, A. ed., 1887. Gli ordinamenti sanitari del comune di Pistoia contro la pestilenzia del 1348. *Archivio Storico Italiano* Ser. 4, *20*, pp.8–22.

Christakos, G., Olea, R.A., Serre, M.L., Wang, L.L. and Yu, H.L., 2005. *Interdisciplinary public health reasoning and epidemic modelling: the case of Black Death,* p. 320. New York: Springer.

Christakos, G., Olea, R.A. and Yu, H.L., 2007. Recent results on the spatiotemporal modelling and comparative analysis of Black Death and bubonic plague epidemics. *Public Health, 121*(9), pp.700–720.

Cipolla, C.M., 1981. *Fighting the plague in seventeenth-century Italy.* University of Wisconsin Press.

Cliff, A.D., Smallman-Raynor, M.R. and Steven, P.M., 2009. Controlling the geographical spread of infectious disease: plague in Italy, 1347–1851. *Acta Medico-Historica Adriatica, 7*(2), pp.197–236.

Cohn, S.K., 2002. *The Black Death transformed: disease and culture in early renaissance Europe.* London: Arnold.

Cohn, S.K., 2007. After the Black Death: labour legislation and attitudes towards labour in late-medieval western Europe. *The Economic History Review, 60*(3), pp.457–485.

Cohn Jr., S.K., 2008. Epidemiology of the Black Death and successive waves of plague. *Medical History Supplement,* (27), pp.74–100.

Cohn, S.K., 2012. Pandemics: waves of disease, waves of hate from the Plague of Athens to AIDS. *Historical Research, 85*(230), pp.535–555.

Cohn Jr., S.K. and Alfani, G., 2007. Households and plague in early modern Italy. *The Journal of Interdisciplinary History, 38*(2), pp.177–205.

Cook, R.A., 1964. The influence of the Black Death on medieval literature and language. *Kentucky Foreign Language Quarterly*, *11*(1), pp.5–13.

Creighton, C.A., 1894. *From the extinction of plague to the present time* (Volume II). Cambridge, UK: Cambridge University Press.

Cromley, E.K., 2010. Pandemic disease in Russia: from Black Death to AIDS. *Eurasian Geography and Economics*, *51*(2), pp.184–202.

Cui, Y., Yu, C., Yan, Y., Li, D., Li, Y., Jombart, T., Weinert, L.A., Wang, Z., Guo, Z., Xu, L. and Zhang, Y., 2013. Historical variations in mutation rate in an epidemic pathogen, Yersinia pestis. *Proceedings of the National Academy of Sciences*, *110*(2), pp.577–582.

Cummins, N., 2014. Longevity and the rise of the west: lifespans of the European Elite, 800–1800. Available at SSRN 2496939.

Curtis, D.R., 2016. Was plague an exclusively urban phenomenon? Plague mortality in the seventeenth-century low countries. *Journal of Interdisciplinary History*, *47*(2), pp.139–170.

Davis, D.E., 1986. The scarcity of rats and the Black Death: an ecological history. *Journal of Interdisciplinary History*, *16*, pp.455–470.

DesOrmeaux, A.L., 2007. *The Black Death and its effect on fourteenth and fifteenth century art*. LSU Master's Thesis.

Devaux, C.A., 2013. Small oversights that led to the great plague of Marseille (1720–1723): lessons from the past. *Infection, Genetics and Evolution*, *14*, pp.169–185.

DeWitte, S.N. and Wood, J.W., 2008. Selectivity of Black Death mortality with respect to preexisting health. *Proceedings of the National Academy of Sciences*, *105*(5), pp.1436–1441.

Dragic, T., Litwin, V. and Allaway, G.P., 1996. HIV-1 entry into CD4 plus cells is mediated by the chemokine receptor CC-CKR-5. *Nature*, *381*(6584), p.667.

Drancourt, M. and Raoult, D., 2002. Molecular insights into the history of plague. *Microbes and Infection*, *4*(1), pp.105–109.

Drancourt, M., Roux, V., Tran-Hung, L., Castex, D., Chenal-Francisque, V., Ogata, H., Fournier, P.E., Crubézy, E. and Raoult, D., 2004. Genotyping, Orientalis-like Yersinia pestis, and plague pandemics. *Emerging Infectious Diseases*, *10*(9), pp.1585–1592.

Duncan, C.J. and Scott, S., 2005. What caused the Black Death? *Postgraduate Medical Journal*, *81*(955), pp.315–320.

Duncan, S.R., Scott, S. and Duncan, C.J., 2005. Reappraisal of the historical selective pressures for the CCR5-Δ32 mutation. *Journal of Medical Genetics*, *42*(3), pp.205–208.

Dyer, C., 2003. Introduction to R. Hilton. *Bond men made free: medieval peasant movements and the English rising of 1381*. Abingdon, UK: Routledge.

Eckert, E.A., 1996. *The structure of plagues and pestilences in early modern Europe. Central Europe, 1560–1640*. Basel, Switzerland: Karger.

Eckert, E.A., 2000. The retreat of plague from Central Europe, 1640–1720: a geomedical approach. *Bulletin of the History of Medicine*, *74*(1), pp.1–28.

EINITE. www.dondena.unibocconi.it/wps/wcm/connect/cdr/centro_dondena/home/research/einite, accessed November 18, 2016.

Eiríksson, J., Bartels-Jonsdottir, H.B., Cage, A.G., Gudmundsdottir, E.R., Klitgaard-Kristensen, D., Marret, F., Rodrigues, T., Abrantes, F., Austin, W.E., Jiang, H. and Knudsen, K.L., 2006. Variability of the North Atlantic Current during the last 2000 years based on shelf bottom water and sea surface temperatures along an open ocean/shallow marine transect in western Europe. *The Holocene*, *16*(7), pp.1017–1029.

Ell, S.R., 1989. Three days in October of 1630: detailed examination of mortality during an early modern plague epidemic in Venice. *Reviews of Infectious Diseases*, *11*(1), pp.128–139.

Epstein, S.R., 2002. *Freedom and growth: the rise of states and markets in Europe, 1300–1750* (Volume 17). Abingdon, UK: Routledge.

Frati, P., 1999. Quarantine, trade and health policies in Ragusa-Dubrovnik until the age of George Armmenius-Baglivi. *Medicina nei secoli, 12*(1), pp.103–127.

Gage, K.L. and Kosoy, M.Y., 2005. Natural history of plague: perspectives from more than a century of research. *Annual Review of Entomology, 50*, pp.505–528.

Galvani, A.P. and Slatkin, M., 2003. Evaluating plague and smallpox as historical selective pressures for the CCR5-Δ32 HIV-resistance allele. *Proceedings of the National Academy of Sciences, 100*(25), pp.15276–15279.

Goldberg, P.J.P., 1988. Women in fifteenth-century town life. In *Towns and townspeople in the fifteenth century* (pp.107–128), ed. J.A.F. Thomson. Gloucester, UK: Sutton Publishing.

Haddock, D.D. and Kiesling, L., 2002. The Black Death and property rights. *The Journal of Legal Studies, 31*(S2), pp.S545–S587.

Haensch, S., Bianucci, R., Signoli, M., Rajerison, M., Schultz, M., Kacki, S., Vermunt, M., Weston, D.A., Hurst, D., Achtman, M. and Carniel, E., 2010. Distinct clones of Yersinia pestis caused the Black Death. *PLoS Pathogens, 6*(10), p.e1001134.

Hall, A., 2008. *Plague in London: a case study of the biological and social pressures exerted by 300 years of Yersinia pestis*. Doctoral dissertation, Oregon State University.

Harding, V., 2002. *The dead and the living in Paris and London, 1500–1670*. Cambridge, UK: Cambridge University Press.

Herlihy, D. 1997. *The Black Death and the transformation of the West*. Cambridge, USA: Harvard University Press.

Hickey, T.M., 2014. *Arrival from abroad: plague, quarantine, and concepts of contagion in eighteenth-century England*. History Undergraduate Thesis, University of Washington, Paper 6.

Hoffman, P.T., Jacks, D.S., Levin, P.A. and Lindert, P.H., 2002. Real inequality in Europe since 1500. *The Journal of Economic History, 62*(2), pp.322–355.

Hopkins, S.V., 1956. Seven centuries of the prices of consumables, compared with builders' wage-rates. *Economica, 23*(92), pp.296–314.

Karlsson, G., 1996. Plague without rats: the case of fifteenth-century Iceland. *Journal of Medieval History, 22*(3), pp.263–284.

Kelly, J., 2006. *The great mortality: an intimate history of the Black Death*. New York: Harper Collins.

Kitsikopoulos, H., 2002. The impact of the Black Death on peasant economy in England, 1350–1500. *The Journal of Peasant Studies, 29*(2), pp.71–90.

Konstantinidou, K., Mantadakis, E., Falagas, M.E., Sardi, T. and Samonis, G., 2009. Venetian rule and control of plague epidemics on the Ionian Islands during 17th and 18th centuries. *Emerging Infectious Diseases, 15*(1), p.39.

Kool, J.L. and Weinstein, R.A., 2005. Risk of person-to-person transmission of pneumonic plague. *Clinical Infectious Diseases, 40*(8), pp.1166–1172.

Liu, X. and Shaffer, L., 2007. *Connections across Eurasia: transportation, communication, and cultural exchange on the Silk Roads*. New York: McGraw-Hill.

Manfredini, M., De Iasio, S. and Lucchetti, E., 2002. The plague of 1630 in the territory of Parma: outbreak and effects of a crisis. *International Journal of Anthropology, 17*(1), pp.41–57.

Martine, G. and McGranahan, G., 2013. The legacy of inequality and negligence in Brazil's unfinished urban transition: lessons for other developing regions. *International Journal of Urban Sustainable Development, 5*(1), pp.7–24.

Martinson, J.J., Chapman, N.H., Rees, D.C., Liu, Y.T. and Clegg, J.B., 1997. Global distribution of the CCR5 gene 32-basepair deletion. *Nature Genetics, 16*(1), pp.100–103.

Marwick, J.D., 1574. Extracts from the records: 1574, Oct–Dec. In *Extracts from the records of the Burgh of Edinburgh, 1573–1589* (pp. 27–34), ed. J.D. Marwick (Edinburgh, 1882). *British History Online.* www.british-history.ac.uk/edinburgh-burgh-records/1573-89/pp27-34, accessed August 18, 2016.

Mate, M.E., 1998. *Daughters, wives, and widows after the Black Death: women in Sussex, 1350–1535.* Suffolk, UK: Boydell & Brewer.

Mead, R., 1722. *A short discourse concerning pestilential contagion and the methods to be used to prevent it.* London: Buckley.

Mecsas, J., Franklin, G., Kuziel, W.A., Brubaker, R.R., Falkow, S. and Mosier, D.E., 2004. Evolutionary genetics: CCR5 mutation and plague protection. *Nature, 427*(6975), p.606.

Melikishvili, A., 2006. Genesis of the anti-plague system: the Tsarist period. *Critical Reviews in Microbiology, 32*(1), pp.19–31.

Morea, V., 1817. *Storia della peste di Noja* (488 p.). Naples: Tipografia Di Angelo Trani.

Mullan, J., 2007. *Mortality, gender, and the plague of 1361–2 on the estate of the bishop of Winchester.* Cardiff School of History and Archaeology.

Naphy, W.G., 2002. *Plagues, poisons, and potions: plague-spreading conspiracies in the western Alps, c. 1530–1640.* Manchester: Manchester University Press.

Noble, J.V., 1974. Geographic and temporal development of plagues. *Nature, 250*(5469), pp.726–729.

Nükhet, V., 2015. *Plague and empire in the early modern Mediterranean world: the Ottoman experience, 1347–1600.* New York: Cambridge University Press.

Olea, R.A. and Christakos, G., 2005. Duration of urban mortality for the 14th-century Black Death epidemic. *Human Biology, 77*(3), pp.291–303.

Orent, W., 2004. *Plague: the mysterious past and terrifying future of the world's most dangerous disease.* New York: Simon and Schuster.

Palmer, D., 1998. Plague. In *Infectious diseases* (2nd edn, pp. 1568–1575), eds. S L. Gorbach, J. G. Bartlett and N. R. Blacklow. Philadelphia, PA: Saunders.

Palmer, R.J., 1978. *The control of plague in Venice and Northern Italy 1348–1600.* Doctoral dissertation, University of Kent.

Pamuk, Ş., 2007. The Black Death and the origins of the "great divergence" across Europe, 1300–1600. *European Review of Economic History, 11*(3), pp.289–317.

Pouqueville, F.C.H., 1805. *Voyage en morée, a Constantinople, en Albanie: et dans plusieurs autres parties de l'Empire Othoman, pendant les annés 1798, 1799, 1800 et 1801 comprenant la description de ces pays* (Volume 2, pp. 109–111). Paris: Troisième tome.

Pruitt, S., 2014. Medieval "Black Death" was airborne, scientists say. www.history.com/news/medieval-black-death-was-airborne-scientists-say, accessed online November 8, 2016.

Rezakhani, K., 2010. The road that never was: the Silk Road and Trans-Eurasian exchange. *Comparative Studies of South Asia, Africa and the Middle East, 30*(3), pp.420–433.

Rielly, K., 2010. The black rat. In *Extinctions and invasions. A social history of British fauna* (pp. 134–145), eds. T. O'Connor and N. Sykes. Oxford, UK: Windgather Press.

Rigby, S.H., 2000. Gendering the Black Death: women in later medieval England. *Gender & History, 12*(3), pp.745–754.

Righi, A., 1633. *Historiae contagiosi morbi qui florentiam populatus fuit anno 1630.* Florence: Typis Francisci Honufrij.

Rothenberg, G.E., 1973. The Austrian sanitary cordon and the control of the bubonic plague: 1710–1871. *Journal of the History of Medicine and Allied Sciences*, *28*(1), pp.15–23.

Schmid, B.V., Büntgen, U., Easterday, W.R., Ginzler, C., Walløe, L., Bramanti, B. and Stenseth, N.C., 2015. Climate-driven introduction of the Black Death and successive plague reintroductions into Europe. *Proceedings of the National Academy of Sciences*, *112*(10), pp.3020–3025.

Schuenemann, V.J., Bos, K., DeWitte, S., Schmedes, S., Jamieson, J., Mittnik, A., Forrest, S., Coombes, B.K., Wood, J.W., Earn, D.J. and White, W., 2011. Targeted enrichment of ancient pathogens yielding the pPCP1 plasmid of Yersinia pestis from victims of the Black Death. *Proceedings of the National Academy of Sciences*, *108*(38), pp.E746–E752.

Scott, S. and Duncan, C., 2001. *Biology of plagues: evidence from historical populations*. Cambridge, UK: Cambridge University Press.

Seifert, L., Wiechmann, I., Harbeck, M., Thomas, A., Grupe, G., Projahn, M., Scholz, H.C. and Riehm, J.M., 2016. Genotyping Yersinia pestis in historical plague: evidence for long-term persistence of Y. pestis in Europe from the 14th to the 17th century. *PLoS One*, *11*(1), p.e0145194.

Signoli, M., Séguy, I., Biraben, J.N., Dutour, O. and Belle, P., 2002. Paleodemography and historical demography in the context of an epidemic: plague in provence in the eighteenth century. *Population (English Edition)*, *57*, pp.829–854.

Slack, P., 1988. *Poverty and policy in Tudor and Stuart England*. Essex, UK: Longman.

Sloan, A.W., 1973. Medical and social aspects of the great plague of London in 1665. *South African Medical Journal*, *47*(7), pp.270–276.

Snow, J., 1854. The cholera near Golden-square, and at Deptford. *Medical Times Gazette*, *9*, pp.321–322.

Spyrou, M.A., Tukhbatova, R.I., Feldman, M., Drath, J., Kacki, S., de Heredia, J.B., Arnold, S., Sitdikov, A.G., Castex, D., Wahl, J., Gazimzyanov, I.R., Nurgaliev, D., Herbig, A., Bos, K. and Krause, J., 2016. Historical Y. pestis genomes reveal the European Black Death as the source of ancient and modern plague pandemics. *Cell Host & Microbe*, *19*(6), pp.874–881.

Steckel, R.H., 2004. New light on the "dark ages": the remarkably tall stature of northern European men during the Medieval era. *Social Science History*, *28*(2), pp.211–228.

Stenseth, N.C., Samia, N.I., Viljugrein, H., Kausrud, K.L., Begon, M., Davis, S., Leirs, H., Dubyanskiy, V.M., Esper, J., Ageyev, V.S. and Klassovskiy, N.L., 2006. Plague dynamics are driven by climate variation. *Proceedings of the National Academy of Sciences*, *103*(35), pp.13110–13115.

Van Damme, W. and Van Lerberghe, W., 2000. Editorial: epidemics and fear. *Tropical Medicine and International Health*, *5*(8), pp.511–514.

Vazquez-Prokopec, G.M., Bisanzio, D., Stoddard, S.T., Paz-Soldan, V., Morrison, A.C., Elder, J.P., Ramirez-Paredes, J., Halsey, E.S., Kochel, T.J., Scott, T.W. and Kitron, U., 2013. Using GPS technology to quantify human mobility, dynamic contacts and infectious disease dynamics in a resource-poor urban environment. *PloS One*, *8*(4), p.e58802.

Velimirovic, B. and Velimirovic, H., 1989. Plague in Vienna. *Review of Infectious Diseases*, *11*(5), pp.808–826.

Voigtländer, N. and Voth, H.J., 2012. The three horsemen of riches: plague, war, and urbanization in early modern Europe. *Review of Economic Studies*, *80*(2), pp.774–811.

Walløe, L., 2008. Medieval and modern bubonic plague: some clinical continuities. *Medical History*, *52*(S27), pp.59–73.

Watts, S., 2001. Some aspects of mortality in three Shropshire parishes in the mid-seventeenth century. *Local Population Studies, 67*, pp.11–25.

Welford, M.R. and Bossak, B.H. 2009. Validation of inverse seasonal peak mortality in medieval plagues, including the Black Death, in comparison to modern Yersinia pestis-variant Diseases. *PLoS One, 4*(12), p.e8401.

Whittles, L.K. and Didelot, X., 2016. May. Epidemiological analysis of the Eyam plague outbreak of 1665–1666. In *Proceedings of Royal Society B, 283*(1830), p.20160618.

Wilschut, L.I., Addink, E.A., Heesterbeek, H., Heier, L., Laudisoit, A., Begon, M., Davis, S., Dubyanskiy, V.M., Burdelov, L.A. and de Jong, S.M., 2013. Potential corridors and barriers for plague spread in central Asia. *International Journal of Health Geographics, 12*(1), p.49.

Wilschut, L.I., Laudisoit, A., Hughes, N.K., Addink, E.A., Jong, S.M., Heesterbeek, H.A., Reijniers, J., Eagle, S., Dubyanskiy, V.M. and Begon, M., 2015. Spatial distribution patterns of plague hosts: point pattern analysis of the burrows of great gerbils in Kazakhstan. *Journal of Biogeography, 42*(7), pp.1281–1292.

Wood, J. and DeWitte-Aviña, S., 2003. Was the Black Death yersinial plague? *The Lancet Infectious Diseases, 3*(6), pp.327–328.

Xu, L., Stige, L.C., Kausrud, K.L., Ari, T.B., Wang, S., Fang, X., Schmid, B.V., Liu, Q., Stenseth, N.C. and Zhang, Z., 2014. Wet climate and transportation routes accelerate spread of human plague. *Proceedings of the Royal Society of London B: Biological Sciences, 281*(1780), p.20133159.

Yang, Y., Sugimoto, J.D., Halloran, M.E., Basta, N.E., Chao, D.L., Matrajt, L., Potter, G., Kenah, E. and Longini, I.M., 2009. The transmissibility and control of pandemic influenza A (H1N1) virus. *Science, 326*(5953), pp.729–733.

Yue, R.P., Lee, H.F. and Wu, C.Y., 2016. Navigable rivers facilitated the spread and recurrence of plague in pre-industrial Europe. *Scientific Reports, 6*, p.34867.

Zimbler, D.L., Schroeder, J.A., Eddy, J.L. and Lathem, W.W., 2015. Early emergence of Yersinia pestis as a severe respiratory pathogen. *Nature Communications, 6*, p.7487.

6 Re-emergence in China and spread to Singapore, Taiwan, Bombay, San Francisco, and Australia before 1901

Third Plague Pandemic begins

Today, we live in the Third Plague Pandemic; between 1954 and 1997 the World Health Organization (WHO) documented 80,613 cases, in some 38 countries, of which 6,587 people infected with plague died of it, a case fatality rate of 8% (Bramanti *et al.* 2016). Very few plague cases occurred in Europe; in fact, only 7,000 people died of plague between 1899 and 1950 in Europe, while as few as 500 people have died in the United States (Echenberg 2002). Today, most plague deaths are restricted to the Middle East, East, South and South-East Asia, Africa, the Caribbean, and those countries south of the US border. Other than a few outbreaks of pneumonic plague in Manchuria and Madagascar, most cases are bubonic plague.

In fact, today, epidemiologists suggest plague was transmitted from rats (plague strain F1991016 or *biovar orientalis*) to humans as a bubonic plague (Zhou *et al.* 2004; Zimbler *et al.* 2015) somewhere in Yunnan, China in 1772. Plague remained in Yunnan province in western China until the Panthay or Du Wenxiu rebellion, in which the ethnic Muslim Hui peoples rose up against the Qing dynasty between 1856 and 1873 (Curson and McCracken 1989). The social and economic devastation wrought by the Imperial suppression of the rebellion, with some 1 million Hui people being killed, provided the opportunity for the plague to escape the confines of the providence. The capital of Yunnan, K'un-ming, was infected with plague in 1866, where it remained for the next two to three decades (Curson and McCracken 1989). Thereafter, plague reached the south-east coast of China, infecting Beihai in 1867, then Lianzhou, Quingzhou, the Leizhou Peninsula, and Hainan Island before infecting the open port of Guangzhou (Canton) in 1894, killing in excess of 70,000 people (Bramanti *et al.* 2016; Curson and McCracken 1989; Echenberg 2010; Lee 2013). In Canton, as elsewhere in China, the Qing regime was hesitant to implement quarantine measures, even those as simple as isolating the sick, because isolation of the sick from family members violated Confucian codes (Echenberg 2010). As a result of the inconsistent and hesitant application of modern public quarantine health measures by the Qing regime, plague spread unhindered, jumping from port to port along the Chinese coast. Even though the

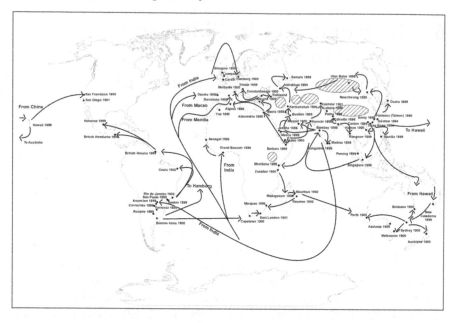

Figure 6.1 The diffusion of the Third Pandemic 1894–1901. Hatching indicates areas of
plague endemism. (Data obtained from P.H. Curson and K. McCracken, 1989.
Plague in Sydney: The anatomy of an epidemic. UNSW Press.)

Chinese Imperial government established quarantine measures in Shanghai and
Xiamen in 1873, they quarantined Beihai in 1877, a decade after it first erupted
in Beihai in 1867 (Echenberg 2010; Lee 2013). This delay was repeated along the
Chinese coast with Shantou quarantined in 1883, Ningbo in 1894, Tianjin in 1895,
Liuzhuan in 1899, Fuzhou in 1900, Hankou in 1902, and Guangzhou in 1911,
some 15 years after Guangzhou was first infected with plague (Lee 2013).

Once bubonic plague reached Guangzhou (Canton) in 1894, the chances of
containing and quarantining the plague and plague victims were slim to none. In
fact, some 40,000 Hong Kong Chinese residents attended the Chinese New Year
celebrations in Canton in 1894 (Echenberg 2010) and in doing so, probably contributed to plague's emergence in Hong Kong that year. However, the likely route
of plague's emergence in Hong Kong was through maritime trade; for instance, in
1894 approximately 30% of all Chinese exports, or 22 million tons, and 41% of all
Chinese imports moved through Hong Kong (Echenberg 2010; Keller *et al.* 2011).
Within a decade of plague's arrival in Hong Kong, British steamships moving
goods, people, and rats across the British Empire facilitated its spread around the
world (Figure 6.1) (Echenberg 2002).

From the port of Hong Kong, plague spread initially south and west; Macao
and Fuzhou were infected in 1895 and Singapore, Taiwan, and Bombay in 1896
(Curson and McCracken 1989). Taiwan subsequently lost 760 to the plague

(Curson and McCracken 1989). Plague also spread eastwards from Hong Kong, infecting Kobe, Japan in 1898, San Francisco in March 1899, and Manila and Australia in 1900 (Curson and McCracken 1989; Echenberg 2002, 2010; Eldridge 1900; Lee 2013; Peckham 2013).

Just as the news reports of the eruption of Krakatoa in August 26–27, 1883 illustrated the beginning of a global news media coverage (Winchester 2004), so the spread of the bubonic plague from Hong Kong through the UK's colonial trading network was charted in real time by the fledgling global news media (Peckham 2013). By the late 1880s, telegraph cables from London to Hong Kong meant important news could circumnavigate the globe in 53 minutes (Baark 1997; Peckham 2013). These two events and their media coverage were seminal events in the globalization of information and news, heralding a new smaller world linked by capital (Mackinder 1899) and news media. As the age of geographic discovery closed in the late 1890s so the age of global interdependence (the fourth great transition) began; anything local had the potential to become global, be it news, goods, money, or disease (Mackinder 1899; McMichael 2004; Peckham 2013; Watts and Strogatz 1998). Since the 1890s, modern transportation systems have become perfect vectors for newly emergent microbes, be they viral or bacterial (Bossak and Welford 2010; Watts and Strogatz 1998).

However, plague occurred in Cutch, India in 1812, killing 50% of the population there before spreading to Gujarat, Sind and Ahmedabad, while in the 1830s 20% of Pali died of plague (Klein 1988); thankfully, a strict quarantine prevented this plague epidemic from spreading beyond Pali. Nevertheless, plague was present in the Himalayan foothills from the 1830s–1850s and spilled onto the Gangetic plain in the early 1850s, killing thousands (Klein 1988). The failure of these localized plague epidemics to erupt across India prior to 1896 might be due to this plague being sustained by a plague strain that afflicts the Indian gerbil, *Tatera indica*, rather than the plague (plague strain F1991016 or *biovar orientalis*) that was transmitted by rat fleas and sustained the Third Plague Pandemic from Hong Kong to Bombay in August of 1896.

Bombay, another major trading port, was just as crucial to the globalization of plague as the port of Hong Kong was initially. From Bombay, plague spread west and south to east Africa, Madagascar, and Mauritius, where 1,691 people died between 1899 and 1900, and north-west into the Red Sea (Curson and McCracken 1989). Jeddah, in modern-day Saudi Arabia, was first infected in early 1896, but a full-blown epidemic did not affect Jeddah, Mecca, and Medina until 1899 (Curson and McCracken 1989). From the port of Yanbu, which acts as the entry point for Muslim pilgrims to Mecca and Medina, plague spread to North Africa, infecting Alexandria, Egypt, on May 4, 1899, where between May 20 and November 2, 1899, 45 people died of plague (Long 1900). Thereafter, Constantinople was infected in August 1900 (Curson and McCracken 1989). It appears that pilgrims or the boats leaving Yanbu also infected Kamaran Island, South Yemen, Aden, and Berbera in 1899, while another wave of plague followed pilgrims to Muscat and Basra in May 1899 (Curson and McCracken 1989).

From Bombay, another stream of plague moving southwards infected Tamatav, Madagascar by 1898, Mozambique by 1899, Réunion by 1899, and King Williams Town near Cape Town, South Africa by November 1900 (Curson and McCracken 1989; Eager 1908; Mitchell 1983). Thereafter, the plague reached Europe, with Lisbon and Astrakhan infected in 1899, and Glasgow by 1900 (Eager 1908). In Russia, fishing ports around the Caspian Sea were infected by 1899 (Curson and McCracken 1989).

In Central and South America, Santos, Brazil, and La Boca, Panama suffered plague epidemics in 1899, while Asunción was infected in September 1899, and Buenos Aires and Rosario were infected in January 1900 (Eager 1908). Santos was probably infected via a ship out of Oporto, while Asunción was probably infected by a grain ship originating in India (Curson and McCracken 1989).

Epidemiology and re-emergence of plague in China

The re-emergence of plague in China and its arrival in Hong Kong triggered, in today's CDC lexicon, several 'outbreak response teams' being sent to the epidemic. In this case, teams from the Pasteur Institute in France (invited by the Hong Kong government) and Japan (invited by the British Foreign Office) descended on Hong Kong (Collins 1999; Lee 2013). The Japanese team, headed by Aoyama Tanemichi and Kitasato Shinasaburŏ, was worried about the plague spreading to Japan and wanted to be the first to isolate the pathogen responsible for the plague epidemic, while the Japanese government wanted to cultivate a political friendship with Hong Kong and the British government (Lee 2013; Tadashi 1994). Interestingly, the Japanese team was given extensive help, use of medical facilities, and access to plague victims; in contrast, the French team was denied access to plague victims and had to build their own facilities, and only with the help of an Italian missionary were the French able to access any plague victims (Collins 1999; Lee 2013). This hostility towards the French was probably driven by the British Colonial Office concerns over French expansion in Asia (Bibel and Chen 1976; Lee 2013).

In Hong Kong in July, 1894, both Alexander Yersin and Kitasato Shinasaburŏ, head of a Japanese medical delegation, isolated the plague bacterium, although it was Alexander Yersin who was credited with the discovery of the pathogen responsible for the Third Plague Pandemic partly because Yersin produced a more accurate description of the bacterium (Butler 1983; Lee 2013; Peckham 2016). Alexander Yersin also noted that the streets of Hong Kong were littered with tens of thousands of dead rats (Bramanti *et al.* 2016). In 1898, in a controlled laboratory test, Paul-Louis Simond showed that the lesions on Karachi residents' legs and feet were in fact the location of rat flea bites and that plague was spread by the bite of rat fleas harboring the bacteria, *Yersinia pestis*, in their foregut (Bramanti *et al.* 2016). Simond went further and showed that once a rat corpse cooled below a specific temperature, fleas would abandon their host rat and seek a blood-meal elsewhere (Bramanti *et al.* 2016). On occasion, in crowded slums or during cold winters, these blood-meals would turn out to be humans (Keeling and Gilligan 2000).

Culture, politics and plague in Hong Kong and Bombay/Mumbai, India

Colonial Hong Kong in 1894 was a territory riven by racism (Peckham 2016), class-based distrust, and physician rivalries (Choa 1993) that undermined the ability of the colony to quarantine the plague. Contemporary commentators and colonial administrators conflated 'Chinese things' and 'junk' (or rubbish) as being likely sources of plague infection (Peckham 2016). In doing so, Europeans sought to blame the Chinese for the plague and its lethality. Beginning in the 1850s, western medical literature pathologized the Chinese, just as British colonial officials did with Indians after the Sepoy Rebellion of 1857 (Paxman 2011; Peckham 2016). These efforts saw Europeans viewing both India and China as pathological bodies (Peckham 2016). Once Hong Kong was declared an infected port on May 10, 1894, sanitation teams and 300 men from the Shropshire regiment began house-to-house inspections and disinfections; ~7000 of tons of personal belongings, furniture and other items were removed from tenements in the city or burned or disinfected with carbolic acid in the streets (Peckham 2016). It was not until 1897–1898 that work published by Walter Wyman, Paul Louis Simond and William Simpson condemned this as unnecessary and showed that rats were the conveyors of plague (Bramanti *et al.* 2016; Peckham 2016).

Aggressive anti-plague measures effected by the British in May 1894 quickly precipitated violent responses from the Hong Kong Chinese on May 19 and 20, as draconian measures targeted ethnic Chinese rather than the minority white British (Echenberg 2010). Hong Kong Governor Robinson reacted by reducing plague measures in late May and allowing repatriation of sick ethnic Chinese to mainland China, starting on June 9 (Echenberg 2010). This effectively broke the quarantine, and plague was once more free to spread further afield. Not for the first time, nor the last, was public health put aside for politics (Echenberg 2010).

Racism was not the only issue to confront those facing the plague epidemic in Hong Kong; the British Colonial Office was also more concerned with plague's potential to disrupt local and global imperial trade, further limit port incomes, and reduce labor through death, illness, and plague-induced flight panic (Peckham 2013). It did not help global public health that the governor of Hong Kong eased quarantine restrictions in 1894 (Peckham 2013). In 1898, the British extracted from the Chinese a lease for the New Territories of Kowloon as a means to control the flight of Chinese labor, again illustrating that the British Colonial Office saw the Hong Kong plague in terms of an economic crisis (Echenberg 2010; Peckham 2013). Fear of the plague and subsequent port quarantines around the world contributed to a series of global trade recessions (Peckham 2013).

Hong Kong was poorly prepared to handle the arrival of plague in May 1894; it had neither a chief medical officer nor any death-certificate legislation, and only two government-run hospitals, so it is hardly surprisingly that the epidemic lasted 30 years, with 21,867 cases, 20,489 fatalities and a 93.7% case fatality rate (Choa 1993; Echenberg 2010). Plague also arrived in Hong Kong at a time of acute labor exploitation, when approximately a third of all laborers were bonded to contractors and not able to sell their labor independently (Echenberg 2010). As a consequence, many Chinese Hong Kong laborers were vulnerable, deeply impoverished, malnourished, and desperate, subsisting on a few bowls of rice a day (Echenberg 2010). As a community, these Chinese laborers suffered poor health and weakened immune systems, all of which probably contributed to plague's high mortality in Hong Kong.

Within mainland China, the arrival of plague in the Pearly River estuary that connects Canton with Hong Kong undermined the Qing dynasty (Echenberg 2010). Foreign powers insisted the Qing dynasty enact quarantine and travel restrictions, yet the failure of the dynasty to do this systematically, and the loss of honor accorded the dynasty as it became seen as a puppet of western governments, inspired many radicals to consider overthrowing the Qing dynasty as the only option to save China (Echenberg 2010).

Similarly, the failure of the British to contain rat-borne plague in India once it arrived in Bombay in August 1896 is subject to much debate and recrimination. Although the Indian Medical Service established a whole suite of anti-plague measures, which included mandatory hospitalization of all plague sufferers, quarantine of all people who had contact with known people infected with plague, evacuation of places that had plague, disinfection of plague victims' homes, examination of all travelers and banning of all religious pilgrimages, plague swept India (Klein 1988). This failure was the result of distrust and fear among the Indian population of the British Raj (or British Colonial System) that administered the Indian Medical Service. The savage reprisals that the East India Company inflicted on Indians following the Sepoy Uprising or Indian Rebellion of 1857–1858 cast a deep shadow over British-Indian relations that undermined all subsequent British-sponsored humanitarian efforts (Paxman 2011). As a result, the East India Company was dissolved, and the British government took over. The Sepoy Uprising also increased the high degree of cultural autonomy between the ruling British and the colonized, disenfranchised Indian populace (Arnold 1986). Prior to the rebellion, fraternization in the form of cohabitation among the British (employees of the East India Company) and Indian population was ignored; in fact, many British men took Indian wives and/or girlfriends, who occupied a respected place in society (Paxman 2011). Thereafter cohabitation was banned by the British government and, as a result, the British found themselves increasingly isolated from the Indian population.

The increased isolation of the British Raj from the Indian populace was also replicated in the imposition of public health care policy and activities; the goal of public health reform in British India was to just protect the British in India from what the British Raj felt were "medically dangerous hosts" (Echenberg 2010, p. 54). The British-centric nature of public health in India was nurtured by notions that 'cleansing' India for Indians was impossible and cost-prohibitive, while increased local municipality and regional autonomy effectively stymied expensive responses to plague threats, and the bureaucratic dogma and inflexibility of the Indian Medical Service (IMS) meant doctors focused their attention on British military and civilian needs (Echenberg 2010).

In Bombay in 1896, intense anti-plague measures failed to immediately halt the advance of plague. These precipitated violent backlashes against the IMS, hospitals and staff, and the British Colonial officers who were trying to enforce these anti-plague measures (Klein 1988). Bombay's position as British India's principal port and railroad nexus on its west coast effectively sealed India's fate; arriving in Bombay on September 23, 1896, the plague spread rapidly among the impoverished, overcrowded city blocks, infecting and killing thousands before advancing into Bombay's hinterland and the Indian interior, as Indians fled by train and on foot after the anti-plague measures were lifted (Echenberg 2010). The violent backlashes to British organized anti-plague measures in the 1890s also contributed to the rise in Indian nationalism (Echenberg 2002).

Speed, seasonality, duration and mortality of the Third Plague Pandemic in China and India

Roads, rivers and coastlines all increased plague velocities, while landscape ruggedness reduced plague spread velocities, yet wet years increased plague velocities (Xu *et al.* 2014). Although plague leapt rapidly between continents on ocean-going transport, within continents the Third Plague Pandemic moved much, much slower than mBD velocities. The medieval Black Death was transmitted north and east in Europe at an average rate of 0.9 to 6 km a day (Benedictow 2004; Christakos *et al.* 2005, 2007) "compared with BP [bubonic plague] velocities in [the] USA averaging roughly 25 km/year since 1900 and 13–19 km/year between 1866 and 1944 in China" (Benedict 1996). The transmission of mBD was so rapid and the disease was so lethal that it does not fit the typical velocity of diffusion in observed plague epidemics such as China, where plague began in Yunnan in 1772 and reached Hong Kong in May 1894 (Bramanti *et al.* 2016; Choa 1993; Xu *et al.* 2014). In India between 1896 and 1906, the regional diffusion of plague was controlled by grain transports along railroads while, locally, plague failed to cross streets because some houses were more rat-proof than others (Yu and Christakos 2006).

The incidence of plague in India is seasonal and frequently dictated to by the transport of agricultural products (Sharif 1951). Cooler temperatures, higher humidities and the monsoon season rains drove rats inside shelters such as burrows, houses, and warehouses, precipitating renewed outbreaks of plague, while grain (i.e., rice, cotton seed) storage, refining, and transportation centers became the loci of plague in India (Buxton 1932; Hirst 1927; Sharif 1951). Near Bombay, plague first erupted in commercial agricultural centers and then radiated out to surrounding farms, grain-storage depots, and villages (Sharif 1951). Cooler temperatures and higher humidities seemed to encourage rat fleas to multiply and plague to survive within fleas (Sharif 1951). In India, cases of plague are highest between October and February, appearing most strongly correlated with higher saturated humidity (Rogers 1928). Plague also erupted more frequently in Northern India due to the region's cooler temperatures, higher humidities, and larger undernourished populations living in higher densities than Southern India (Klein 1988).

Today, spatial meta-population analyses confirm these early ideas and have established that where plague first arrives in a rat population with a low 25–50% susceptibility, it rarely erupts into the human population (Christakos *et al.* 2007; Keeling and Gilligan 2000). However, if rat susceptibility to plague is greater than 80%, rats suffer a lethal, rather short-lived epidemic that increases the likelihood of plague transmission to humans (Keeling and Gilligan 2000). High rat susceptibility to plague is both temporal and spatial; rat populations that have not been exposed to plague for many rat generations exhibit higher susceptibility, and these tend to either be geographically isolated in some way or simply distant from any endemic plague areas. In India, these areas are typically hot, dry areas away from rice-producing locations (Christakos *et al.* 2007; Sharif 1951). The incidence of plague today and in the past, is also a function of urban rodent composition. Mumbai (Bombay) suffers high incidence of plague because 65–70% of its rodent population are black rats, and 49–77% of these rats have *X. chepis* (plague-carrying fleas) (Klein 1988). In contrast, Calcutta has a lower incidence of plague because its major rodent species is a bandicoot that harbors far fewer *X. chepis* fleas (Klein 1988).

In China prior to 1901, plague infected 2.6 million people, killing 2.2 million of them (Bramanti *et al.* 2016; Xu *et al.* 2014). Spatial analyses of georeferenced plague data from China between 1772 and 1964 indicate that plague spread at median rates across all of China that varied between 11.8 and 40.1 km/year (Xu *et al.* 2014). Comparison of the medieval Black Death (mBD) and the Indian plague of 1900 (± 10 years) using spatial modeling in GIS identified the following significant inconsistencies: mBD mortality was two orders higher than in India; modern bubonic plague is a rural disease, whereas many isolated rural villages, particularly in north-west Spain, were bypassed by the mBD (Benedictow 2004; Biraben 1975); and mBD transmission velocities were higher than observed in India (Yu and Christakos 2006; Christakos *et al.* 2007). In contrast to the initial episode or primary wave of the medieval Black Death, now known to have been pneumonic plague, that swept across Europe and western Russia from 1347–1352, the bubonic plague in India lasted for six uninterrupted decades beginning in 1896, with peak mortality in 1908, and receding to near extinction by 1940 (Christakos *et al.* 2007; Seal 1960). In its first decade in India, plague killed 6 million people, increasing the annual death rate to 4.32% (Seal 1960). In comparison, the mBD killed at least a third of all Europeans between 1347 and 1353 (Benedictow 2004). Moreover, the medieval Black Death killed most Europeans in the spring through late summer when commingling of people in fairs, markets, and religious pilgrimages was at the peak, whereas the Indian Plague was most lethal in winter when rats and people retreated indoors (Welford and Bossak 2009). In China, data from 1850–1964 suggest that the effect of precipitation on plague eruption is non-linear (Xu *et al.* 2011). In the drier north, wet years increased plague pandemics, but these were scattered and non-continuous across the landscape, although extremely wet years decreased plague incidence (Xu *et al.* 2011). In contrast, in the wetter, more humid south, increased precipitation actually reduced plague incidence, but where plague pandemics erupted these were continuous across space (Xu *et al.* 2011).

Bubonic plague's low case fatality rate

The arrival of plague in China, India, and beyond, occurred rather fortuitously as modern scientific, laboratory-based medicine, revolutionized by rapid developments during two world wars (e.g., the discovery of and commercial development of penicillin), became globally available (Echenberg 2002). But just as scientific, laboratory-based medicine came of age in the early part of the 20th century, so too did public health (Echenberg 2002). Today, both India and China, two rapidly emerging economies, rarely play host to the plague. Their accumulated wealth, relatively well-funded public health systems, and widespread access to low-cost antibiotics have greatly reduced the incidence of plague. Sadly, inequalities in access to medicine and public health facilities across the rest of the Global South (e.g., clean drinking water, sewage disposal) remain (Echenberg 2002). Although plague case fatality rates have remained low since the 1950s, the Global South suffers more plague epidemics than the richer Global North.

Impact of plague on daily life during the plague

Benedictow (2004) claims "plague normally arrived with persons unwittingly carrying infective rat fleas in their clothing or luggage" (Cohn 2002). In fact, Indian plague commissioners found no evidence that clothing and luggage from people emigrating from plague-stricken regions dispersed *Y. pestis* (Cohn 2002; Gatacre 1897). Instead, in India the spread of plague mirrored the transport of rat-favored merchandise (e.g., wheat) by railway tracks, motor lorries and bullock-carts (Sharif 1951). In contrast to the mBD, where family members infected other family members of a household, in India, plague commissioners found only 4% of households with more than one infected family member (Cohn 2002; Gatacre 1897). Again, in contrast to the mBD where hospital workers were more susceptible to plague than others, the plague commissioners found that "the safest place to be in plague time is the plague ward" (Cohn 2002; Gatacre 1897; Kool and Weinstein 2005). Yet between 1872 and 1921, life expectancies fell 20 to 25 years as plague ravaged India, killing 12 million people (Klein 1988). Plague terrorized people and almost led to the collapse of Bombay's Punjab economies, as millions evacuated their homes to camp in fields, while millions of others were detained in observation camps (Klein 1988).

Although black rats and their fleas appear to be the main vector of plague-to-human transmission in India, it appears that plague is endemic to the Indian gerbil, *Tatera indica* (Klein 1988). Plague eruptions among rats and subsequently humans occur when plague erupts among gerbil colonies that exist in close proximity to humans and their attendant rats (Klein 1988). The intersection of gerbils and humans appears to have several foci in India – the Himalayan foothills, the

Vindhya, Bhanrer and Maikal ranges in Central India, the Western Ghats, and Hyderabad (Seal 1960). Furthermore, human vulnerability to plague increased during periods of nutrient shortage, in other words, during famines (Griffiths 1983; Klein 1988). In India, the widespread presence of malaria and diarrhea increased the likelihood of plague case fatalities; however, wealth and caste membership also affected plague mortality (Klein 1988). In 1897, in Porbander, higher castes and wealthy families suffered far fewer plague case fatalities than the lower castes and the poor: just two of the financial elites died, while 340 manual laborers died, among them beggars, unskilled laborers, butchers, and weavers (Klein 1988). The wealth bias in plague mortality witnessed in the late 1890s in India is likely due to the greater exposure of the poor to rat fleas, a higher incidence of malnutrition, and their immune systems being overburdened with many chronic debilitating diseases such as diarrhea and malaria (Klein 1988). Elsewhere in India, grain dealers, due to their close proximity to rats, suffered higher than typical mortalities, as did Jains, who are traditionally merchants (Klein 1988). However, sustained grain famines seem to mirror reductions in rat populations and reductions in plague deaths among humans in India (Klein 1988).

Both India and China were in the late 1890s–1920s vulnerable to plague and remain so today, because both were and remain a mix of modernity and underdevelopment. Both have modern transportation systems and particularly extensive rail networks that foster a high degree of human and rodent mobility, yet both exhibit extreme and extensive poverty, where the poor live in crowded unsanitary conditions and exposure to rat fleas is high, and both exhibit high rates of malnutrition and immune deficiencies (Klein 1988). Moreover, India's vast commerce and huge grain trade coupled with the transport of rice, bajri, and wheat to famine-affected regions across India in the late 1890s provided an efficient means to disperse plague-infected rats and their fleas across India (Christakos *et al.* 2007; Klein 1988). Furthermore, at the beginning of the Third Plague Pandemic, much of India and China's populations had had very little exposure to plague (compared with Europeans) and so exhibited little individual and/or limited herd immunity to plague (Klein 1988). Both countries, particularly northern India and west, central, and northern China, offered favorable climates for sustaining rat fleas (Klein 1988).

Plague control in China, India and elsewhere prior to 1900

Rapid industrialization and growth and global integration of the Indian Ocean Rim, and Pacific Rim cities of Macao, Fuzhou, Singapore, Taiwan, Kobe, Manila, Bombay, San Francisco, and Australia, coupled with significant Chinese diaspora (from the mid-19th century to 1930, Kuhn 2006) and Indian diaspora (from 1834–1920 to British, Dutch and French colonies, Pandey *et al.* 2006), facilitated the spread of plague out of south-east Asia and beyond (Echenberg 2002). British colonial office concerns about economic downturns in and around 1900 outweighed any public health concerns; thus regional and ultimately global plague control was ineffective at best (Peckham 2013).

In the late 1890s and 1900s, local plague control and rat eradication programs were also ineffective and haphazard. Because bubonic plague is a zoonosis associated with the transfer of plague from rats via their fleas to humans, vaccination of humans does not eradicate the disease, nor do rat eradication programs that start once the human plague epidemic is underway (Keeling and Gilligan 2000). In fact, rat eradication might increase human epidemics as rat fleas seek new blood-meal sources once their host rat has been killed (Keeling and Gilligan 2000). Therefore, urban rat eradication programs are only effective prior to rats suffering any plague epidemic or if these rat eradication programs keep rat population densities below 3,000 rats/km^2 (Keeling and Gilligan 2000), something urban governments worldwide struggle to achieve. Moreover, the presence of endemic plague-tolerant rodents across many continents provides urban rats with a continuous source of plague, and in some cases, new antibiotic-resistant strains of plague (Keeling and Gilligan 2000). Killing rat fleas through the application of pesticide in known rat areas and the rat-proofing of houses, commercial seed warehouses and transportation of seeds or grains has proved the most successful means to limit rat-known plague epidemics worldwide (Borchert *et al.* 2012; Brygoo 1966; Conrad *et al.* 1968; Curson and McCracken 1989; Eskey 1934).

The spread of plague away from China and India

Plague in Brazil in 1899

Plague arrived in Brazil in São Paulo on October 18, 1899, causing 39 cases and 15 deaths by February 1900, when effective quarantine measures ended the epidemic (Echenberg 2010). However, Rio was infected in mid-April 1900, and by March 1901 plague had infected 599 people, killing 304, with the poorest neighborhoods hit hardest after significant rat die-offs (Echenberg 2010). Plague, though, did not leave; between 1900 and 1950, Brazil suffered 9,532 cases, most in 1903 and 1904, with 819 and 1,035 cases (Echenberg 2010).

Plague in Japan in 1899

Kobe and Osaka were first infected with plague in November 1899, while the plague epidemic in Osaka, across the bay from Kobe, ended in and around January 11, 1900 (Eldridge 1900). Plague erupted again in Osaka, either on April 8 or 10, but abated July 16 (Eldridge 1900). This information was obtained first-hand by Stuart Eldridge, Acting Assistant Surgeon of the US Marine-Hospital Service and part of the maritime quarantine service stationed at Yokohama (Eldridge 1900). In reality, Eldridge was a Sanitary Inspector and part of the US Pacific Rim disease early warning system. He was responsible for inspecting all ships and their passengers, crew and goods that were sailing through Yokohama, Kobe, and Osaka from China or Japan to the US (Barde 2003). Thereafter, Osaka experienced a small outbreak of 14 cases with eight fatalities between September 11 and September 14 (Eldridge 1900).

Eldridge was convinced that the plague epidemics that gripped Kobe and Osaka from November 1899 to July 16, 1900 arrived from Yokohama and, before that, China (Barde 2003). However, subsequent plagues seemed to arrive from the east. On August 15, 1900, a Chinese steerage passenger en route from Honolulu to Kobe on board the steamer *Coptic*, which originally sailed from San Francisco, was hospitalized in Kobe with the plague (Eldridge 1900). The passenger, a rice farmer living 14 miles from Honolulu, exhibited signs of both bubonic and pneumonic plague and died on August 18 (Eldridge 1900). At the time of embarkation, Honolulu was free of plague but San Francisco was not, so it would appear that plague came from San Francisco, where plague killed 110 people in 1900 (Barde 2003; Eldridge 1900).

Plague in Australia in 1900

Arriving in 1900, plague broke out in Sydney with 303 cases and 103 fatalities (Curson and McCracken 1989). The first case occurred on January 19, 1900 when a carter (or porter) came down with plague, while the peak of the epidemic occurred between April 28 and May 21 with 30 deaths, and concluded August 18 with the last plague death (Cumpston and McCallum 1926). The first plague-infected rats were identified February 14, on a wharf in Sydney harbor (Cumpston and McCallum 1926). Although Brisbane was put on alert, the first plague-infected black rats were found dead on March 5 in a shop adjacent to the wharf

Figure 6.2 'Slum' buildings in the Rocks and Paddington areas including Millers
Point were demolished in 1901 as part of the plague cleansing operations.
(Reprinted with permission from State Library of NSW archives and John
Degotardi Jnr collection.)

at which a vessel from Sydney had just berthed, although the first human case in Brisbane was not established until April 27 (Cumpston and McCallum 1926). The fear of plague in Sydney led to a quarantine from March 24 to July 17, the clearing of 'slums' (Figure 6.2), and the culling of 34,500 rats, including both the native bush rat, which does not carry plague, and the introduced black rat (Banks *et al.* 2011). Today, over 100 years later, efforts are still ongoing to try and restore Sydney's foreshore wildlife through the reintroduction of bush rats (Banks *et al.* 2011). Between 1900 and 1908, in contrast to mBD's seasonal peak in summer (Welford and Bossak 2009), human plague cases peaked in May during the southern hemisphere's winter months (Cumpston and McCallum 1926).

Plague in San Francisco in 1900

Rather infamously and ironically, Eldridge was involved in the San Francisco plague outbreak of 1900 because he certified the *Nippon Maru* free of plague after being quarantined for a week in Nagasaki after a Chinese passenger fell ill with plague on May 26, 1900 just six days after the ship left Hong Kong (Barde 2003). Three days out from Honolulu, another *Nippon Maru* Chinese steerage passenger died of plague (this was later confirmed in Honolulu), and upon arrival in Hawaii, the boat and its seven Honolulu-bound passengers and 244 San Francisco-bound passengers were subjected to another strict quarantine (Barde 2003). The Honolulu passengers were subject to a seven-day quarantine, while the San Francisco passengers were subject to a four-day quarantine (Barde 2003). Thereafter, at 1:30 p.m. on June 22, 1900 the *Nippon Maru* continued to San Francisco with its 2,500 tons of freight (Barde 2003) and *plague-infected rats*. Although strict quarantines of all passengers and crew of the *Nippon Maru* were enforced in both Yokohama and Honolulu, no attempt was made to fumigate the cargo, even though passengers and crew and their belongings were disinfected (Eldridge 1900). This is surprising given that Eldridge was aware that Japanese plague researchers were already convinced that the most important transmitters of plague were rats (Eldridge 1900).

Pneumonic plague in Manchuria

Although bubonic plague swept through China and on to India in the 1890s, pneumonic plague ravaged Manchuria; in 1910–1911 killing ~60,000, in 1917–1918 killing ~16,000, and again in 1920–1921 killing 9,000 (Bramanti *et al.* 2016; Kool and Weinstein 2005; Nishiura *et al.* 2006). During this time period, Manchuria was under Japanese imperial control. In both the 1910–1911 and 1920–1921 epidemics, epidemic velocities were very high, with plague moving 965 km in just three weeks in 1910 (Chermin 1989). The 1910–1911 pneumonic plague is believed to have originated among marmot hunters but thereafter was sustained among poor train travelers who rode in poorly vented, overcrowded railway carriages (Kool and Weinstein 2005). If DNA material can be obtained from these Manchurian epidemics confirming the source of plague as marmot plague, we might have escaped a global apocalypse. Let us not forget that we have abundant

DNA evidence that implicates marmots as the source of the plague that caused the medieval Black Death (Bos *et al.* 2011; Spyrou *et al.* 2016).

Social consequences of plague prior to 1901

Just as in previous centuries and plague outbreaks, plague changed the political, economic, social, and cultural landscapes of the countries it infested. In this case, it was the countries of the Indian Ocean and Pacific Rims of Asia. Plague sowed the seeds for the Chinese revolution of 1947, undermined British rule in India, established disease early warning systems, and was a seminal event in communications. The Third Plague also exposed the shortcomings of the sanitarian's view of confronting pandemics, with its emphasis on profiling ethnic minorities, strangers, traders, and travelers (with its inherent racist overtones; McClain 1988), quarantine, and disinfecting people, their luggage, and property versus the newer, modern bacteriological/virological view of pathogens that sought to identify pathogenic hosts, vectors, and source areas (Echenberg 2010).

As noted earlier, Europeans, in particular the British, in India, China and Hong Kong, and later, Americans in San Francisco (McClain 1988), came to view Indians and Chinese as pathological bodies (Peckham 2016). During the San Francisco 1900 plague epidemic, the Surgeon General of the United States quarantined the entire Chinese quarter on San Francisco after the body of Chick Gin was discovered at 1001 Dupont Street (Kalisch 1972; McClain 1988). Even after repealing the quarantine, the city's Chinese population were threatened with inoculation with the dangerous and not necessarily effective Haffkine's prophylactic vaccine (McClain 1988). This approach was successfully fought in the courts, but a second quarantine of Chinese in San Francisco was enacted but overturned in the law courts (McClain 1988). All these measures failed to constrain plague; in all, 113 people, mostly Chinese, died. Thereafter, plague continued to spread across the western states of the US. Such open racism displayed in San Francisco during the 1900 plague helped in the social construction of the Chinese Other, whereby Americans tried to exclude Chinese who in their eyes did not fit the norm of their social group (Power 1995).

The tension between sanitarian and pathogenic views is also illustrated by Eldridge, Acting Assistant Surgeon of the US Marine-Hospital Service, who as part of the maritime quarantine service stationed at Yokohama in 1900, ignored recommendations from Japanese researchers that the source of plague was rats and rat fleas rather humans or their luggage. Many plague inspectors and plague quarantine doctors remained convinced that quarantine was the only way forth to fight plague because quarantine had been so effective in the cessation of the Second Plague Pandemic. But plague had changed; rather than being a human-to-human transmitted pneumonic form of plague that persisted throughout the Second Plague Pandemic and hence necessitated quarantine of humans, plague during the current Third Plague Pandemic is bubonic. A non-human host, be it a rat or other mammal, is necessary to act as a host for bubonic plague before plague is communicated to humans who act as pathogenic dead-ends typically incapable of infecting other humans (Keeling and Gilligan 2000; Stenseth *et al.* 2008). As a result, human quarantine of bubonic plague is ineffective (Keeling and Gilligan 2000); instead public health officers must concentrate on eliminating the source of the plague – rats or other mammals carrying plague-vectoring fleas. During the 1900 San Francisco plague epidemic, no systematic attempt was made to deal with rats. Only in 1902 did San Francisco hire rat-catchers and in 1904 begin rat-proofing buildings (McClain 1988). In contrast, in Sydney, Australia in 1900, rat eradication was prioritized (Curson and McCracken 1989).

References

Arnold, D., 1986. Cholera and colonialism in British India. *Past & Present, 113*, pp.118–151.

Baark, E., 1997. *Lightning wires: the telegraph and China's technological modernization 1860–1890*. Westport, CT, USA: ABC-CLIO/Greenwood.

Banks, P., Cleary, G. and Dickman, C., 2011. Sydney's bubonic plague outbreak 1900–1910: a disaster for foreshore wildlife? *Australian Zoologist, 35*(4), pp.1033–1039.

Barde, R., 2003. Prelude to the plague: public health and politics at America's Pacific Gateway, 1899. *Journal of the History of Medicine and Allied Sciences, 58*(2), pp.153–186.

Benedict, C., 1996. *Bubonic plague in nineteenth-century China*. Palo Alto, CA: Stanford University Press.

Benedictow, O.J., 2004. *The Black Death, 1346–1353: the complete history*. Suffolk, UK: Boydell & Brewer.

Bibel, D.J. and Chen, T.H., 1976. Diagnosis of plaque: an analysis of the Yersin-Kitasato controversy. *Bacteriological Reviews, 40*(3), p.633.

Biraben, J.-N., 1975. *Les hommes et al peste en France et dans les pays europeens et mediterraneens*. Volumes 1 (455 p.) & 2 (416 p.). Paris, France: Mouton.

Borchert, J.N., Eisen, R.J., Atiku, L.A., Delorey, M.J., Mpanga, J.T., Babi, N., Enscore, R.E. and Gage, K.L., 2012. Efficacy of indoor residual spraying using lambda-cyhalothrin for controlling nontarget vector fleas (Siphonaptera) on commensal rats in a plague endemic region of northwestern Uganda. *Journal of Medical Entomology, 49*(5), pp.1027–1034.

Bos, K.I., Schuenemann, V.J., Golding, G.B., Burbano, H.A., Waglechner, N., Coombes, B.K., McPhee, J.B., DeWitte, S.N., Meyer, M., Schmedes, S. and Wood, J., 2011. A draft genome of Yersinia pestis from victims of the Black Death. *Nature, 478*(7370), pp.506–510.

Bossak, B.H. and Welford, M.R., 2010. Spatio-temporal attributes of pandemic and epidemic diseases. *Geography Compass, 4*(8), pp.1084–1096.

Bramanti, B., Stenseth, N.C., Walløe, L. and Lei, X., 2016. Plague: a disease which changed the path of human civilization. In *Yersinia pestis: retrospective and perspective* (pp. 1–26), eds. R.Yang and A. Anisimov, Netherlands: Springer.

Brygoo, E.R., 1966. Epidémiologie de la peste à Madagascar. *Les Archives de l'Institut Pasteur de Madagascar, 35*(1), pp.9–147.

Butler, T.C., 1983. *Plague and other Yersinia infections*. Plenum Medical 24.

Buxton, P.A., 1932. The climate in which the rat-flea lives. *Indian Journal of Medical Research, 20*(1), pp.281–297.

Chernin, E., 1989. Richard Pearson Strong and the Manchurian epidemic of pneumonic plague, 1910–1911. *Journal of the History of Medicine and Allied Sciences, 44*, pp.296–319.

Choa, G.H., 1993. The Lowson diary: a record of the early phase of the Hong Kong Bubonic plague 1894. *Journal of the Hong Kong Branch of the Royal Asiatic Society*, *33*, pp.129–145.

Christakos, G., Olea, R.A., Serre, M.L., Wang, L.L. and Yu, H.L., 2005. *Interdisciplinary public health reasoning and epidemic modelling: the case of Black Death* (p. 320). New York: Springer.

Christakos, G., Olea, R.A. and Yu, H.L., 2007. Recent results on the spatiotemporal modelling and comparative analysis of Black Death and bubonic plague epidemics. *Public Health*, *121*(9), pp.700–720.

Cohn, Jr, S.K., 2002. The Black Death: end of a paradigm. *The American Historical Review*, *107*(3), pp.703–738.

Collins, R.J., 1999. The Black Death: Hong Kong 1894. Lecture, Hong Kong Museum of Medical Sciences, April 24.

Conrad, F.G., LeCocq, F.R. and Krain, R., 1968. A recent epidemic of plague in Vietnam. *Archives of Internal Medicine*, *122*(3), pp.193–198.

Cumpston, J.H.L. and McCallum, F., 1926. The history of plague in Australia, 1900–1925. Commonwealth of Australia: Department of Health, Service Publication No. 32.

Curson, P.H. and McCracken, K., 1989. *Plague in Sydney: the anatomy of an epidemic*. Kensington, Australia: UNSW Press.

Eager, J.M., 1908. *The present pandemic of plague* (No. 22). Wasington DC: Government Printing Office.

Echenberg, M.J., 2002. Pestis redux: the initial years of the third bubonic plague pandemic, 1894–1901. *Journal of World History*, *13*(2), pp.429–449.

Echenberg, M.J., 2010. *Plague ports: the global urban impact of bubonic plague, 1894–1901*. New York: NYU Press.

Eldridge, S., 1900. Plague in Japan from July 1 to September 15, 1900. *Public Health Reports (1896–1970)*, pp.2717–2719.

Eskey, C.R., 1934. Epidemiological study of plague in the Hawaiian Islands. *Public Health Bulletin* 213.

Gatacre, W.F., 1897. *Report: bubonic plague in Bombay for 1896–97*. India: Times of India Steam Press.

Griffiths, E., 1983. Adaptation and multiplication of bacteria in host tissues. *Philosophical Transactions of the Royal Society of London B: Biological Sciences*, *303*(1114), pp.85–96.

Hirst, L.F., 1927. Rat-flea surveys and their use as a guide to plague preventive measures. *Transactions of the Royal Society of Tropical Medicine and Hygiene*, *21*(2), pp.87–108.

Kalisch, P.A., 1972. The Black Death in Chinatown: plague and politics in San Francisco 1900–1904. *Arizona and the West*, *14*(2), pp.113–136.

Keeling, M.J. and Gilligan, C.A., 2000. Metapopulation dynamics of bubonic plague. *Nature*, *407*(6806), pp.903–906.

Keller, W., Li, B. and Shiue, C.H., 2011. China's foreign trade: perspectives from the past 150 years. *The World Economy*, *34*(6), pp.853–892.

Klein, I., 1988. Plague, policy and popular unrest in British India. *Modern Asian Studies*, *22*(4), pp.723–755.

Kool, J.L. and Weinstein, R.A., 2005. Risk of person-to-person transmission of pneumonic plague. *Clinical Infectious Diseases*, *40*(8), pp.1166–1172.

Kuhn, P.A., 2006. Why China historians should study the Chinese diaspora, and vice-versa. *Journal of Chinese Overseas*, *20*(2), pp.163–172.

Lee, P.T., 2013. Colonialism versus Nationalism: the plague of Hong Kong in 1894. *The Journal of Northeast Asian History*, *10*(1), pp.97–128.

Long, J.C., 1900. Egypt. Plague in Alexandria in 1899. *Public Health Reports*, *15*(26), pp.1662–1665.

Mackinder, H., 1899. The great trade routes: lecture V to the Institute of Bankers. UK.

McClain, C., 1988. Of medicine, race, and American law: the bubonic plague outbreak of 1900. *Law & Social Inquiry*, *13*(3), pp.447–513.

McMichael, A.J., 2004. Environmental and social influences on emerging infectious diseases: past, present and future. *Philosophical Transactions of the Royal Society of London B: Biological Sciences*, *359*(1447), pp.1049–1058.

Mitchell, F.K., 1983. The plague in Cape Town in 1901 and its subsequent establishment as an endemic disease in South Africa. *South African Medical Journal*, *63*(27), pp.17–19.

Nishiura, H., Schwehm, M., Kakehashi, M. and Eichner, M., 2006. Transmission potential of primary pneumonic plague: time in homogeneous evaluation based on historical documents of the transmission network. *Journal of Epidemiology & Community Health*, *60*(7), pp.640–645.

Pandey, A., Aggarwal, A., Devane, R. and Kuznetsov, Y., 2006. The Indian diaspora: a unique case? In *Diaspora networks and the international migration of skills* (pp.71–97), ed. Y. Kuznetsov. WBI Development Studies.

Paxman, J., 2011. *Empire: what ruling the world did to the British*. London: Viking.

Peckham, R., 2013. Infective economies: empire, panic and the business of disease. *The Journal of Imperial and Commonwealth History*, *41*(2), pp.211–237.

Peckham, R., 2016. Hong Kong Junk: plague and the economy of Chinese things. *Bulletin of the History of Medicine*, *90*(1), pp.32–60.

Power, J.G., 1995. Media dependency, bubonic plague, and the social construction of the Chinese other. *Journal of Communication Inquiry*, *19*(1), pp.89–110.

Rogers, L., 1928. The yearly variations in plague in India in relation to climate: forecasting epidemics. *Proceedings of the Royal Society of London. Series B, Containing Papers of a Biological Character*, *103*(721), pp.42–72.

Seal, S.C., 1960. Epidemiological studies of plague in India: 1. The present position. *Bulletin of the World Health Organization*, *23*(2–3), p.283.

Sharif, M., 1951. Spread of plague in the southern and central divisions of Bombay province and plague endemic centres in the Indo-Pakistan subcontinent. *Bulletin of the World Health Organization*, *4*(1), p.75.

Spyrou, M.A., Tukhbatova, R.I., Feldman, M., Drath, J., Kacki, S., de Heredia, J.B., Arnold, S., Sitdikov, A.G., Castex, D., Wahl, J. and Gazimzyanov, I.R., 2016. Historical Y. pestis genomes reveal the European Black Death as the source of ancient and modern plague pandemics. *Cell Host & Microbe*, *19*(6), pp.874–881.

Stenseth, N.C., Atshabar, B.B., Begon, M., Belmain, S.R., Bertherat, E., Carniel, E., Gage, K.L., Leirs, H. and Rahalison, L., 2008. Plague: past, present, and future. *PLoS Medicine*, *5*(1), p.e3.

Tadashi, O., 1994. *Kindai Nihon no taigai senden*. Tokyo: Kenbun Shuppan.

Watts, D.J. and Strogatz, S.H., 1998. Collective dynamics of "small-world" networks. *Nature*, *393*(6684), p.440.

Welford, M.R. and Bossak, B.H., 2009. Validation of inverse seasonal peak mortality in medieval plagues, including the Black Death, in comparison to modern Yersinia pestis-variant diseases. *PLoS One*, *4*(12), p.e8401.

Winchester, S., 2004. *Krakatoa: the day the world exploded*. London, UK: Penguin.

Xu, L., Liu, Q., Stige, L.C., Ari, T.B., Fang, X., Chan, K.S., Wang, S., Stenseth, N.C. and Zhang, Z., 2011. Nonlinear effect of climate on plague during the third pandemic in China. *Proceedings of the National Academy of Sciences*, *108*(25), pp.10214–10219.

Xu, L., Stige, L.C., Kausrud, K.L., Ari, T.B., Wang, S., Fang, X., Schmid, B.V., Liu, Q., Stenseth, N.C. and Zhang, Z., 2014. Wet climate and transportation routes accelerate spread of human plague. *Proceedings of the Royal Society of London B: Biological Sciences*, *281*(1780), p.20133159.

Yu, H.L. and Christakos, G., 2006. Spatiotemporal modelling and mapping of the bubonic plague epidemic in India. *International Journal of Health Geographics*, *5*(1), p.12.

Zhou, D., Han, Y., Song, Y., Huang, P. and Yang, R., 2004. Comparative and evolutionary genomics of Yersinia pestis. *Microbes and Infection*, *6*(13), pp.1226–1234.

Zimbler, D.L., Schroeder, J.A., Eddy, J.L. and Lathem, W.W., 2015. Early emergence of Yersinia pestis as a severe respiratory pathogen. *Nature Communications*, *6*, p.7487.

7   1901 to present

Third Plague Pandemic continues

Since the turn of the 20th century, the safest place to be in a plague or other disease epidemic is a hospital ward (Cohn 2008) because wards are typically free of rats, and few people pass either human saliva or blood between patients and nurses or patients and next-of-kin. However, there are exceptions, such as Ebola wards in Sierra Leone and Liberia. So although we live in the Third Plague Pandemic, the likelihood of contracting plague has decreased drastically since 1346. It would appear that human-to-human transmission of plague through pneumonic plague is a thing of the past, but the question remains as to why. Did plague somewhere and sometime between 1346 and 1879 evolve to be a fully vascular (blood-borne) pathogen rather than a pathogen restricted to an infection of the lungs (Cui *et al.* 2013)? Or did bubonic plague outcompete pneumonic plague to become the dominant plague type? Did the source of highly virulent, highly transmissible pneumonic plague cease to exist (Bos *et al.* 2011), or has it failed recently to cross over into humans? Biological samples from skeletal material obtained from plague victims in Europe stretching nearly the length of the Second Pandemic (1348–1710) identify a new (previously uncharacterized) plague genome (Bos *et al.* 2011). This finding suggests that the medieval Black Death that afflicted Europe was caused by a unique and now, quite possibly, extinct plague strain that originated among marmots (Bos *et al.* 2011). In fact, Tumanskii (1957, 1958) and Levi (*et al.* 1961) have identified four plague strains: a rat-borne, marmot-borne, suslik-borne, and vole-borne (Gage and Kosoy 2005). Of these, the marmot-borne strain (or *medievalis biovar*; Drancourt *et al.* 2004) is the most genetically distinct as it lacks a I259T variant of the protease Pla and thus remains a pneumonic disease (Cui *et al.* 2013; Zimbler *et al.* 2015). The I259T variant optimizes protease Pla activity, allowing plague to become bubonic plague, an invasive vascular infection. Gorshkov *et al.* (2000) identified 85 naturally occurring strains with the greatest diversity present in Asia, while North America harbored the least variety. The lack of genetic diversity in the North America rat-borne *Y. pestis* strain is due to plague's recent introduction into North America (Gage and Kosoy 2005).

In an interesting twist, Rasmussen *et al.* (2015) argue that plague acquired the *ymt* gene that facilitates plague transmission by flea before 951 BCE and that plague was endemic, possibly constrained to local and/or regional populations but not pandemic among all Eurasian human populations between 2,800 and 5,000 years ago. This suggests plague has gone full circle and that pandemic pneumonic plague, or rather the medieval Black Death that raged 1346–1859 (see Chapters 4 and 5), was a very lethal pandemic aberration in an era otherwise dominated by epidemic bubonic plague. From a virological perspective, this makes complete sense; typically pathogens that infect hosts locally and fail to jump scales to a global pandemic stage select for lower virulence, otherwise they kill all available hosts (Boots and Mealor 2007), whereas those pathogens that undergo long-distance dispersal select for higher virulence (Boots and Mealor 2007). This appears to be the case. Mortality rates among untreated bubonic plague patients today range between 50-90%, while today mortality rates for untreated pneumonic plague range between 95-100% (Perry and Fetherston 1997).

The difference in mortality is even more pronounced when one looks at mortality estimates across historic and current plagues. It is generally accepted that 30-40% of Europe's population died between 1346 and 1353 (Benedictow 2004). The high mortalities of this 6- to 7-year period possibly reflect the recent long-distance dispersal of the first strain of marmot-derived pneumonic plague/medieval Black Death (Bos *et al.* 2011; Spyrou *et al.* 2016). Thereafter, during the 500 years of pandemic pneumonic plague/medieval Black Death that afflicted Europe between 1346 and 1815, plague-related mortalities seem to decline (between 1560 and 1640, plague killed just 8–15% of the Eastern European population; Eckert 2000) but increase again after each new marmot-strain of plague dispersed across Eurasia. For instance, 30% of Provence died between 1720 and 1722 following the fourth marmot strain dispersal (Biraben 1975; Signoli *et al.* 2002).

These oscillations in mortality seem to afflict modern bubonic plague. During the Hong Kong epidemic that started in 1894 and lasted three decades, plague had a case fatality rate of 93.7% (Choa 1993). On the Chinese mainland, 2.3 million of 2.6 million Chinese who contracted plague died between 1772 and 1964, but the majority died before 1900 (Xu *et al.* 2014). In contrast, bubonic plague case fatalities between 1954 and 1997 were only 8% (WHO), yet most of these cases occurred in highly localized areas in the Global South with limited access to modern antibiotics and, in Madagascar's case, where resistant strains of plague to antibiotics occur (Dennis and Hughes 1997; Chanteau *et al.* 2000). These spatial and temporal patterns suggest that bubonic plague was more lethal in its early emergent long-distance dispersal phase in China, Hong Kong, and even India, but thereafter, once it had been constrained to poorer, more isolated areas in countries peripheral to global trade, bubonic plague was selected for lower virulence.

Irrespective of the source or virulence of plague, today we live in the era of bubonic plague. It is quite conceivable that increasingly effective human quarantine practices observed in the 1700s and 1800s effectively caused the marmot-derived, pneumonic plague to become extinct, and since then marmot-borne plague has failed to re-emerge as a totally global pandemic among humans. However, Manchuria in 1910–1911 was afflicted with pneumonic plague that appears to have erupted from among Mongolian marmots (Lien-Teh and Tuck 1913a). Rapid and effective implementation of quarantine by the North Manchurian Plague Prevention Service prevented a pandemic (Lien-Teh and Tuck 1913b). Furthermore, since 1900, pneumonic plague has been tougher to catch. Today, typically only patient-homecare providers, hospital nurses, and a patient's immediate family catch pneumonic plague from an infected individual (Begier *et al.* 2006; Gani and Leach 2004; Kool and Weinstein 2005; Ratsitorahina *et al.* 2000; Seal 1960; Wood *et al.* 2003). However, initial pneumonic plague epidemic velocities were high (and involved multiple cases) in Manchuria during the 1910–1911 and 1920–1921 winters (Chernin 1989). In 1910–1911, plague carriers moved 965 km in the first three weeks in overcrowded and poorly ventilated railway cars (Kool and Weinstein 2005; Nishiura *et al.* 2006). However, this was the exception; plague transmission was low to non-existent in well-ventilated hospital wards during both Manchurian epidemics (Kool and Weinstein 2005). However, an early Russian clinical study by Zabolotny suggests that the 1910–1911 epidemic originated among marmots (Gage and Kosoy 2005; Zabolotny 1915). If this is confirmed, it is deeply worrying; the Second Plague Pandemic has been rather conclusively tied to a strain of plague carried by Siberian marmots (Bos *et al.* 2011). Today, these same marmots and their endemic, highly virulent pneumonic plague can be found in the Daurian enzootic area, an area including Hulun Nur Plateau (Mongolia) and the Manchurian Plain (China) (Benedict 1996; Gage and Kosoy 2005).

In contrast, the full globalization of trade, facilitated initially by the Imperial British colonial and trade system (see Chapter 6), that emerged in the late 1800s provided an efficient vector for bubonic-infected rats to translocate around the world and establish plague among endemic rodent populations in the Global South, where plague eradication programs remain ineffective due to chronic underfunding.

During the Indian Plague of the 1900s, epidemic transmission velocities (mapped using Geographic Information Systems) were low, while between 1900 and 1904 plague mortalities remained below 0.4% per year (Yu and Christakos 2006). In fact, plague epidemic velocities in India between 1896 and 1906 were two orders of magnitude slower than epidemic velocities measured during the primary wave of the medieval Black Death (Benedictow 2004; Bossak and Welford 2015; Christakos *et al.* 2007; Cohn 2008; Yu and Christakos 2006). This is due to the fact that the Indian Plague of 1894 to 1906 was bubonic plague. The epidemic velocity of the primary wave of the medieval Black Death was so fast (1.5–6 km/day) that it moved quicker than the Spanish Flu of 1918–1920 (Christakos *et al.* 2007; Cohn 2008), this at a time (1346–1353 CE) when human travel was undertaken by sailboat, horse, or on foot. The Indian plague of 1896–1906 also re-emerged at many localities (Yu and Christakos 2006), something the primary wave of the medieval Black Death (1346–1353) did not. Instead, the primary wave of the mBD progressed across Europe as a broad front (Bossak and Welford 2015; Christakos *et al.* 2007). The Indian Plague of 1896–1906 appeared to mimic the plague eruptions that occurred in Europe from 1361–1879.

Intense, highly discriminatory anti-plague measures enacted by the British in Bombay in 1896 most certainly contributed to the rise in Indian nationalism (Echenberg 2002). The eruption of plague in San Francisco further pathogenized the Chinese living there and contributed to persistent anti-Asian rhetoric and prejudice in the United States (Echenberg 2002). Elsewhere, the eruption of bubonic plague in South Africa certainly appears to have influenced the formulation of urban native policy in the Cape Colony between 1900 and 1909 (Swanson 1977). In fact, plague contributed in South Africa to the development among white South Africans of "infectious disease as a societal metaphor" (Swanson 1977, p. 387). Whereas in industrial societies, overcrowded slums were associated with class and ethnic differences, such as London's East End in the late 1890s, in colonial societies slums were identified with skin color (Swanson 1977). Slum-dwellers in South Africa in the early 1900s, who were nearly entirely black South Africans, became intrinsically linked to infectious disease, and as quarantine was used to isolate people with infectious diseases, so a 'sanitation syndrome' was concocted within the white South African segregationist ideology to validate the establishment of urban apartheid in South Africa (Swanson 1977).

Today and through the last 100 years or so, plague in the US and Madagascar illustrates many of the modern vagaries of plague: its persistence among rodent reservoirs; its genetic flexibility; its continued ability to transmute from rat-flea-human bubonic plague to human-human transmitted pneumonic plague; and its resistance to eradication even among the very wealthiest of countries. Plague is a survivor, an opportunist, and still a killer.

Plague in the US

The first outbreak of plague in the USA occurred between 1900–1904 centered on the port of San Francisco (Adjemian *et al.* 2007; Link 1955; Lipson 1972). A total of 121 cases were identified with a 93% case fatality rate (Echenberg 2007). This outbreak occurred even though quarantine procedures were in place throughout the USA following the news of an outbreak of plague in December 1899 in Honolulu, Hawaii (Link 1955). For instance, SS *Nippon Maru*, a Japanese vessel, was quarantined in San Francisco Bay in June 1899 following the deaths of four people aboard the boat (Link 1955). Subsequent outbreaks from 1904 to 1925 in San Francisco, Seattle, Oakland, and Los Angeles yielded 494 human cases, and these exhibited a 50% case fatality rate (Adjemian *et al.* 2007; Link 1955). Between 1900 and 2012, 1,006 human cases of plague have been documented in the US, of which 80% are bubonic plague (CDC 2016). This compares with the 1,000–2,000 cases reported to the WHO worldwide each year (CDC 2016). It appears that three independent introductions of plague possibly by ship rats infected local rodent populations in San Francisco in 1900, in Seattle in 1907 and in Los Angeles in 1908 (Adjemian *et al.* 2007).

Since 1925, no cases of pneumonic plague have been reported in the US. However, human-to-human transmission of pneumonic plague did occur in Oakland in 1919 and in Los Angeles in 1924 (Kool and Weinstein 2005). In Oakland, a squirrel hunter infected five to six persons, and then these infected another seven people; of these, six survived and were quarantined (Kellogg 1920; Kool and Weinstein 2005). In Los Angeles, 32 people developed pneumonic plague from two people who shared a house and contracted bubonic plague and then spread plague through pneumonic human-human transmission; however, rapid quarantine of infected persons and the use of gowns and masks by health-care workers stopped any further transmission (Meyer 1961; Kool and Weinstein 2005).

Today, plague is still expanding across the western USA among rodents and prairie dogs at velocities of 80.81 ± 15.28 km/year (Adjemian *et al.* 2007), with human incidence of disease peaking during wet late winters/early springs associated with ENSO phases (Parmenter *et al.* 1999; Ari *et al.* 2010). From 1900–2005, transmission velocities between animal hosts were slowest (decreasing -43 ± 6.7 km/year) in Mediterranean California, the Sierra Nevada, the Eastern Cascades, and the Blue Mountains (Adjemian *et al.* 2007), whereas such velocities increased by 26.95 ± 2.57 km/yr in the Great Plains (Adjemian *et al.* 2007). The spread of the plague among rodents in the USA appears to follow a wave-like form; in some areas this wave slows, yet elsewhere, such as in the Southern Rockies, the wave sped up. It appears that this acceleration was the result of farmers translocating infected prairie dogs across the region to control local prairie dog populations with plague (Adjemian *et al.* 2007). Today, plague is monitored closely by the US government, yet having spread 2,250 km in less than 50 years, plague has not moved out of the western Great Plains (Adjemian *et al.* 2007). However, it does appear that sylvatic plague is endemic to black-tailed prairie dogs across their entire northern Great Plains range, even those areas east of active plague such as the Black Hills of Dakota (Mize and Britten 2016).

Among human cases in the USA, in 2006, five (13%) of the 13 cases identified were rather surprisingly septicemic, the rest bubonic (Butler 2009). Case fatality among the 229 cases of human plague (most were the bubonic form) identified in the USA from 1980 through 1994 was quite high, with 33 deaths, or 14% (Dennis and Hughes 1997). Rather interestingly, the presence of late-lying snow in spring in the western US, associated with ENSO and Pacific Decadal Oscillation events, seem to drive the occurrence of rodent and human plague incidence in the USA (Ari *et al.* 2010; Parmenter *et al.* 1999).

Indirectly, in the western Great Plains of the US, mountain plovers that nest among prairie dog colonies suffer declines when prairie dogs experience plague eruptions. In this case, rather than from contracting plague, mountain plover declines are a result of the loss of prairie dogs and the succession of short-grass prairie in and around prairie dog burrows to a longer-grass prairie, as prairie dog consumption of grasses declines (Augustine *et al.* 2008).

In contrast, of the 892 plague cases identified in India between 1994 and 2003, most were pneumonic plague (Butler 2009). The 1994 plague outbreak in India, first reported in the third week of September in Surat, Western India and the first in 30 years in India, caused widespread panic in India and across the globe; it disrupted local, national and global travel and trade (Mavalankar 1995; Dennis and Hughes 1997; Dennis *et al.* 1999).

Plague in Madagascar

Where once plague had disappeared in the 1930s, Madagascar has seen a re-emergence of this disease such that between 1980 and 1995, 45% of African plague cases were from Madagascar (Chanteau *et al.* 1998; Duplantier *et al.* 2003; WHO 1997). Between 1994 and 2003, 12,270 cases of plague with a case fatality of 8% occurred in Madagascar (Butler 2009). Of these, 97% were bubonic plague transmitted via rat fleas and 3% pneumonic plague (Butler 2009). Among the many factors that have contributed to plague's re-emergence in Madagascar are: cuts in plague surveillance programs, increases in poverty, reductions in human health indices, and increases in deforestation that have increased steppe and grassland habitats suitable to rodent colonization (Duplantier *et al.* 2003). It would also appear that the black rat (*Rattus rattus)* found in Madagascar has been at the epicenter of the re-emergence (Gilabert *et al.* 2007). In the rural, central highlands of Madagascar the black rat is both resistant to plague and also the main reservoir of plague (Tollenaere *et al.* 2011; Andrianaivoarimanana *et al.* 2013). Being both resistant and a carrier of plague among rats is highly unusual; typically rats exposed to plague suffer high mortality rates or die-offs (Andriana-ivoarimanana *et al.* 2013). This suggests that selection pressures to elicit such genetic differences between highly resistant, low-susceptibility rats in the highlands and non-resistant, highly susceptible lowland rats is a very recent phenomenon because black rats were only recently introduced in the last few thousand years, while plague was introduced in 1898 (Tollenaere *et al.* 2011). CCR5 polymorphism among black rats may play a role in plague resistance (Tollenaere *et al.* 2008). In humans, a 32-bit deletion on the CCR5 gene was thought to confer immunity to plague, but recent work suggests it confers immunity to smallpox and HIV-1 (Galvani and Novembre 2005). Among rats, a non-synonymous substitution (H184R) on a critical part of the CCR5 gene may provide some form of immunity to plague for rats; however, the association is not statistically valid (Tollenaere *et al.* 2008). Interestingly, experimental work found that Madagascan male black rats are slightly more resistant to plague, while rats injected with 10^5 bacteria died within 3–5 days, but when injected with less than 5,000 bacteria, rats lived and most had no plague antibodies (Tollenaere *et al.* 2010).

Complicating attempts to control plague in Madagascar is that three flea species (e.g., *Xenopsylla cheopis* – the oriental rat flea, *Synopsyllus fonquerniei* – an endemic flea found above 800 m, and *Pulex irritans* – the human flea) are known to facilitate domestic human-to-human transmission of plague (Ratovonjato *et al.* 2014). Interestingly, the endemic flea (*Synopsyllus fonquerniei*) infests outdoor rats mostly above 800 m, while *Xenopsylla cheopis* infests indoor rats (Kreppel *et al.* 2016). There also appears to be a seasonal dimension to plague infection. This appears to be due to seasonal movements of rats; in low plague transmission seasons rats move to sisal hedges and rice paddies, while in high plague transmission seasons rats move into houses (Rahelinirina *et al.* 2010). These seasonal variations might also be genetic. Microsatellite marking of black rats and subsequent genetic analyses suggest that the highland black rats exhibit two genetically different forms: one found outside and with higher burdens of fleas and plague, and the other found in houses (Gilabert *et al.* 2007). The outside form undergoes significant seasonal population booms-and-busts dependent on forage availability, whereas the indoor rats maintain relatively stable populations as food sources are more consistent and environments remain constant (Gilabert *et al.* 2007). Nevertheless, plague transmission to humans suggests fleas move between these two populations, probably during outdoor population crashes. Furthermore, continued global environmental change might reduce areas suitable for *Synopsyllus fonquerniei* and passively reduce the likelihood of future plague epidemics in Madagascar (Kreppel *et al.* 2016*)*.

To complicate matters for Madagascar and the rest of the world, today, multi-drug resistant plague has been observed in Madagascar (Dennis and Hughes 1997; Chanteau *et al.* 2000), and recent outbreaks of plague have occurred in Madagascan port cities (Vogler *et al.* 2013). The combination is extremely worrying. In fact, two *Yersinia pestis* isolates from the port of Mahajanga were found to be resistant to the two anti-plague drugs of choice – chloramphernicol and ampicillin (Chanteau *et al.* 2000), while in the capital Antananarivo, one *Yersinia pestis* isolate was found to be resistant to tetracycline (Chanteau *et al.* 2000). Even with access to modern medicines, between 1996 and 1998 Madagascar suffered case fatality rates of 20%, or approximately 400 deaths, from 2,000 confirmed plague cases (Chanteau *et al.* 2000). In all, it is thought that of the 38 districts affected by the outbreak, 40% of the population was exposed to plague, while 97.2% of the cases were bubonic plague (Chanteau *et al.* 2000). It is important to remember that ports were the entry point into Europe for plague during the Athenian, Justinianic, and primary wave of the medieval Black Death pandemics, while it appears San Francisco was the entry point for plague into the USA. Between 1991 and 1999, the port city of Mahajanga suffered an outbreak of plague. For instance, between July 1995 and March 1996, 617 clinically suspected cases of plague occurred in Mahajanga, but with ready access to streptomycin, a broad-spectrum antibiotic, case fatality rates were kept low at 8.7% (Boisier *et al.* 1997). Figure 7.1 shows the plague districts, the 800-meter altitude that represents the lower limit of the endemic plague-carrying flea, the plague cases from 2007–2011, and the plague port foci of Majunga.

Whole-genome sequence single-nucleotide polymorphism (SNP) analysis has identified several new phylogentic plague lineages in Mahajanga (Vogler *et al.* 2013). rRNA profiling (ribotyping) suggests three new ribotypes or *Y. pestis* strains, distinct from the pre-1982 ribotype that has been associated with the third global plague pandemic, which originated in the Madagascan highlands (Guiyoule *et al.* 1997). It thus appears that multiple strains likely evolved in the highlands above 800 m and were introduced, established, extinguished, and reintroduced multiple times to the port city of Mahajanga. This suggests that expansions of plague are possible into the lowlands and into ports, thereby increasing the likelihood of plague pathogen evolution, amplification, and dispersal beyond Madagascar (Vogler *et al.* 2013). Given the rapidity with which new plague

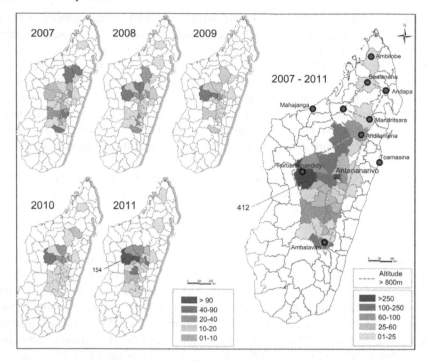

Figure 7.1 Madagascar plague mapping from 2007 to 2011. (Data sources: OCHA, Institut Pasteur de Madagascar. Reprinted from V. Andrianaivoarimanana, K. Kreppel, N. Elissa, J.M. Duplantier, E. Carniel, M. Rajerison and R. Jambou, 2013. Understanding the persistence of plague foci in Madagascar. *PLoS Negl Trop Dis*, *7*(11), p.e2382.)

lineages were generated between 1991 and 1999 in Madajanga, and the presence of drug-resistant strains of plague in the port of Madajanga (sometimes referred to as Majunga), Antananarivo, and Madagascar, it is highly likely that Madagascar will generate and propagate a new lethal, drug-resistant form of plague and disperse this plague worldwide sometime in the future unless more funds are made available to combat plague in Madagascar today. All the more worrying, in 1997 just north of the capital Antananarivo, 18 cases of pneumonic plague were identified, with eight of them dying. This represents the largest outbreak of pneumonic plague in Madagascar since 1957 (Ratsitorahina *et al.* 2000; Kool and Weinstein 2005). The use of the F1-antigen in sputum tests using a simple dipstick provided a sensitive and rapid test for pneumonic plague and proved very helpful in identifying pneumonic plague patients and in containing the epidemic (Ratsitorahina *et al.* 2000). Although efficient, cheap treatment is available to combat plague, misdiagnosis or late diagnosis of plague or the presence of infected persons who do not display clinical manifestations of plague (Ratsitorahina *et al.* 2000) is common in the highlands of Madagascar. Furthermore, limited access to health care, and the lack of trained health-care workers to educate villagers on plague prevention and establish preventive measures to reduce the development of multiple-resistant plague strains all suggest that Madagascar in the late 20th century was a ticking plague time-bomb (Chanteau *et al.* 2000). Few if any health-care workers used masks during the 1991–1999 epidemic, while fewer than 5% of the health-care workers working with plague patients used chemoprophylaxis to reduce their likelihood of infection (Kool and Weinstein 2005). The question, then, was not if but *when* a global drug-resistant plague pandemic would erupt out of Madagascar. Today, the use of inexpensive, widely available, dipstick tests for all forms of plague has, it seems, reduced the plague mortality rate and reduced the incidence of pneumonic plague (Migliani *et al.* 2006), and likely reduced the possibility of a plague pandemic erupting out of Madagascar, but local health-workers and the WHO need to remain watchful.

Plague in Africa since 1901

Since 1877, plague in Africa has erupted along the Mediterranean coast, central, and east African highlands, and either along rivers or coastal areas of southern Africa (Neerinckx *et al.* 2010). All told, 200,000 people have contracted plague in some 26 African countries, and yet plague incidents have increased since 1980, particularly in Tanzania and the Democratic Republic of Congo (DRC) (Davis *et al.* 2006; Neerinckx *et al.* 2010). Today, the Province Orientale of the north-eastern DRC accounts for 50% of all plague cases reported worldwide and 90% of all African cases (Neerinckx *et al.* 2010). It is suspected that the political turmoil, war, civil strife and large refugee populations living in squalid camps in the DRC have had a detrimental impact on wildlife and their habitats, thus seeding many of the plague cases (Dudley *et al.* 2002; Gayer *et al.* 2007; Neerinckx *et al.* 2010).

Just as in Madagascar, the highlands of East and Central Africa remain a major loci of plague (Davis *et al.* 2006; Neerinckx *et al.* 2010), while seasonally plague erupts during dry periods when plague-carrying rodents move out of the fields and into the richer-food environments of villages (Davis *et al.* 2006; Eisen *et al.* 2010). In the Western Usambara Mountains of Tanzania, plague eradication programs have failed to eliminate plague from three remote villages since plague first occurred here in 1980 (Davis *et al.* 2006). Sadly, plague mortalities are 2.2 times higher among 10–14 year olds than 30–34 year olds (Davis *et al.* 2006). In a surprising twist, in the Western Usambara Mountains, plague is transmitted by the human flea, *Pulex irritans*, not a rodent flea. This explains the failure of plague eradication programs that target rodents and their fleas (Laudisoit *et al.* 2007).

Elsewhere in Tanzania, agricultural land supports twice the number of rodents that are plague seropositive than conserved lands (McCauley *et al.* 2015). In this case, it appears that the abundance of the Natal multimammate mouse or African soft-furred rat, *Mastomys natalensis*, in agricultural areas with 44.0 ± 2.0 individuals/ha acts as a plague reservoir (McCauley *et al.* 2015). In contrast, just 2.3 ± 3.2 *Mastomys natalensis* individuals/ha are observed in conserved areas (McCauley *et al.* 2015). This evidence demands that local public health officials across Africa consider the merits of conserving wild lands as a means to reduce future plague outbreaks.

In a frightening replication of the beginning of the highly lethal Manchuria 1910–1911 epidemic that killed in excess of 60,000 people in Uganda in 2004, plague was transmitted by respiratory droplets but not aerosols among four people (Begier *et al.* 2006). The two initial patients infected their caregivers; in one instance, one of the infected individuals was strapped to a bicycle and held upright by his three brothers, who rode 18 km to a medical clinic (Begier *et al.* 2006). He subsequently died, and at his funeral many relatives touched him and the blanket he was wrapped in – all told, some 200 people were exposed to him from his bike ride to the funeral and yet no one else was infected (Begier *et al.* 2006). This case illustrates that bubonic plague is capable, however ineffectively, of seeding human-to-human transmission and that caregivers, be they relatives or medical personnel, must take precautions where plague is suspected.

In summary, it would appear that in Africa, ports, just as in other continents, offer sites for repeated introductions of seropositive plague rodents. But what separates African plague dynamics from other continents is that civil conflicts, war, and escalating deforestation perturb wild lands, triggering movements of seropositive plague rodents into closer contact with at-risk, immune-compromised humans.

Plague in India after 1901

During the first two decades of the Third Plague Pandemic in India that began in 1894, upwards of 25 million people contracted and died of plague at a rate varying between 183 to 133 deaths per million per year, but between 1949–1958 the mortality rate dropped to 1.8 deaths per million per year (Biswas *et al.* 2011). Since 1980, only small, isolated bubonic plague outbreaks have occurred in Himachal Pradesh, Bihar, Karnatak, Maharashtra, Tamil Nadu, Andhra Pradesh, Gujarat, and Uttarakhand (Joshi *et al.* 2009). These outbreaks had low case fatality rates, and most patients suffered only a mild illness, all indicative of a minimally virulent plague strain (John 1995). High rates of seropositivity among the Indian gerbil *Tatera indica cuvieri* suggests the gerbil is the enzootic foci (or wild plague host) in these regions of India (Biswas *et al.* 2011). Peri-domestic rats (*Rattus rattus*) also yield high rates of plague seropositivity, suggesting that these rodents act as go-betweens, moving plague from wild to domestic environments.

In Himachal Pradesh, pneumonic plague erupted in 1983, infecting 22 and killing 17, and in 2002 infecting 30 and killing 5 (Gupta and Sharma 2007; Joshi *et al.* 2009). Again, similar to other recent pneumonic plague outbreaks (e.g., Uganda, Madagascar), only close relatives of the index case in 2002 and nurses administering to those patients were infected. However, an incorrect initial diagnosis of community-acquired pneumonia delayed proper treatment and possibly led to several unnecessary deaths (Gupta and Sharma 2007; Joshi *et al.* 2009).

In Surat, Gujarat on September 23, 1994, 460 pneumonic plague cases were reported by local newspapers (Dutt *et al.* 2006). This received national and worldwide attention, causing over 200,000 Indians and foreigners to flee the states of Gujarat and Maharashtra (John 1994). The sensationalizing of this event by electronic and print media was probably responsible for the panic that ensued across India. The international media in particular were very irresponsible, with several British newspapers evoking scenes from the medieval Black Death when millions were killed (Madan 1995). However, plague abated within a month as health-care workers and government officials worked tirelessly to eradicate it. Those who did not flee retained a high confidence in their local government and public health care. However, where public-health announcements were limited, people resorted to superstitious and unhygienic behavior (Raza *et al.* 1997). From Surat, pneumonic plague diffused across the local region (this is termed 'expansion diffusion'), remaining within 40 km of Surat prior to October 1994, but once people panicked and fled by train from Surat, plague leapfrogged across India in a manner known as 'relocation diffusion' (Dutt *et al.* 2006). Once pneumonic plague had relocated to Delhi, Mumbai and Kolkata, it again diffused locally in a manner referred to as 'hierarchical diffusion', moving from regional centers to local market towns to villages (Dutt *et al.* 2006). Although over 200,000 fled Gujarat and Maharashtra, only 693 plague cases were reported across India, with 56 deaths (CDC 1994).

Plague in Central and South America

As recently as 2010, plague killed one 14-year-old boy and infected another 31 in northern Peru (Daily Mail 2010). Between 1994 and 1999, some 1,700 plague cases were identified among Bolivia, Brazil, Ecuador, Peru, and the US (Ruiz 2001). Plague first arrived in the Americas in Argentina and Brazil in 1899, and by 1906 had dispersed widely across Brazil, becoming resident in a discontinuous manner among native rodents in north-east Brazil (Giles *et al.* 2011). Today, plague still persists and is considered endemic in Bolivia, Brazil, Ecuador, and Peru (Schneider *et al.* 2014). Between 2000 and 2012, 68% of the cases were reported in Peru and 24% in Ecuador, and most cases were associated with high rural poverty and areas above 1,300 m, mirroring both Uganda and Madagascar (Schneider *et al.* 2014). A significant risk factor for contracting plague in the Andes is the presence of domestic guinea pigs inside homes (Schneider *et al.* 2014).

References

Adjemian, J.Z., Foley, P., Gage, K.L. and Foley, J.E., 2007. Initiation and spread of traveling waves of plague, Yersinia pestis, in the western United States. *The American Journal of Tropical Medicine and Hygiene*, 76(2), pp.365–375.

Andrianaivoarimanana, V., Kreppel, K., Elissa, N., Duplantier, J.M., Carniel, E., Rajerison, M. and Jambou, R., 2013. Understanding the persistence of plague foci in Madagascar. *PLoS Neglected Tropical Diseases*, 7(11), p.e2382.

Ari, T.B., Gershunov, A., Tristan, R., Cazelles, B., Gage, K. and Stenseth, N.C., 2010. Interannual variability of human plague occurrence in the Western United States explained by tropical and North Pacific Ocean climate variability. *The American Journal of Tropical Medicine and Hygiene*, *83*(3), pp.624–632.

Augustine, D.J., Dinsmore, S.J., Wunder, M.B., Dreitz, V.J. and Knopf, F.L., 2008. Response of mountain plovers to plague-driven dynamics of black-tailed prairie dog colonies. *Landscape Ecology*, *23*(6), pp.689–697.

Begier, E.M., Asiki, G., Anywaine, Z., Yockey, B., Schriefer, M.E., Aleti, P., Ogen-Odoi, A., Staples, J.E., Sexton, C., Bearden, S.W. and Kool, J.L., 2006. Pneumonic plague cluster, Uganda, 2004. *Emerging Infectious Diseases*, *12*(3), p.460.

Benedict, C.A., 1996. *Bubonic plague in nineteenth-century China*. Palo Alto, CA: Stanford University Press.

Benedictow, O.J., 2004. *The Black Death, 1346–1353: the complete history*. Suffolk, UK: Boydell & Brewer.

Biraben, J.N., 1975. *Les hommes et la peste*. Paris: Mouton.

Biswas, S., Lal, S., Mittal, V., Malini, M. and Kumar, S., 2011. Detection of enzootic plague foci in peninsular India. *Journal of Communicable Disease*, *43*(3), pp.169–176.

Boisier, P., Rasolomaharo, M., Ranaivoson, G., Rasoamanana, B., Rakoto, L., Andrianirina, Z., Andriamahefazafy, B. and Chanteau, S., 1997. Urban epidemic of bubonic plague in Majunga, Madagascar: epidemiological aspects. *Tropical Medicine and International Health*, *2*(5), pp.422–427.

Boots, M. and Mealor, M., 2007. Local interactions select for lower pathogen infectivity. *Science*, *315*(5816), pp.1284–1286.

Bos, K.I., Schuenemann, V.J., Golding, G.B., Burbano, H.A., Waglechner, N., Coombes, B.K., McPhee, J.B., DeWitte, S.N., Meyer, M., Schmedes, S. and Wood, J., 2011. A draft genome of Yersinia pestis from victims of the Black Death. *Nature*, *478*(7370), pp.506–510.

Bossak, B.H. and Welford, M.R., 2015. Spatio-temporal characteristics of the medieval Black Death. *Spatial analysis in health geography, Part II infectious disease* (pp. 71–84). Ashgate's Geographies of Health Series.

Butler, T., 2009. Plague into the 21st century. *Clinical Infectious Diseases*, *49*(5), pp.736–742.

CDC, 1994. Update: human plague—India, 1994. *MMWR Morbidity and Mortal Weekly Report*, *43*(41), pp.761–762.

CDC, 2016. Plague in the United States. www.cdc.gov/plague/maps/, retrieved September 28, 2016.

Chanteau, S., Ratsifasoamanana, L., Rasoamanana, B., Rahalison, L., Randriambelosoa, J., Roux, J. and Rabeson, D., 1998. Plague, a reemerging disease in Madagascar. *Emerging Infectious Diseases*, *4*(1), p.101.

Chanteau, S., Ratsitorahina, M., Rahalison, L., Rasoamanana, B., Chan, F., Boisier, P., Rabeson, D. and Roux, J., 2000. Current epidemiology of human plague in Madagascar. *Microbes and Infection*, *2*(1), pp.25–31.

Chernin, E., 1989. Richard Pearson Strong and the Manchurian epidemic of pneumonic plague, 1910–1911. *Journal of the History of Medicine and Allied Sciences*, *44*(3), pp.296–319.

Choa, G.H., 1993. The Lowson diary: a record of the early phase of the Hong Kong Bubonic plague 1894. *Journal of the Hong Kong Branch of the Royal Asiatic Society*, *33*, pp.129–145.

Christakos, G., Olea, R.A. and Yu, H.L., 2007. Recent results on the spatiotemporal modelling and comparative analysis of Black Death and bubonic plague epidemics. *Public Health*, *121*(9), pp.700–720.

Cohn Jr, S.K., 2008. Epidemiology of the Black Death and successive waves of plague. *Medical History Supplement*, *27*, pp.74–100.

Cui, Y., Yu, C., Yan, Y., Li, D., Li, Y., Jombart, T., Weinert, L.A., Wang, Z., Guo, Z., Xu, L. and Zhang, Y., 2013. Historical variations in mutation rate in an epidemic pathogen, Yersinia pestis. *Proceedings of the National Academy of Sciences*, *110*(2), pp.577–582.

Daily Mail, 2010. Peru teenager killed in bubonic plague outbreak. www.dailymail.co.uk/news/article-1299848/Peru-teenager-killed-bubonic-plague-outbreak.html, accessed July 7, 2017.

Davis, S., Makundi, R.H., Machang'u, R.S. and Leirs, H., 2006. Demographic and spatio-temporal variation in human plague at a persistent focus in Tanzania. *Acta Tropica*, *100*(1), pp.133–141.

Dennis, D.T. and Hughes, J.M., 1997. Multidrug resistance in plague. *New England Journal of Medicine*, *337*(10), pp.702–704.

Dennis, D.T., Gage, K.L., Gratz, N.G., Poland, J.D., Tikhomirov, E. and World Health Organization, 1999. *Plague manual: epidemiology, distribution, surveillance and control*. New York: World Health Organization.

Drancourt, M., Roux, V., Tran-Hung, L., Castex, D., Chenal-Francisque, V., Ogata, H., Fournier, P.E., Crubézy, E. and Raoult, D., 2004. Genotyping, Orientalis-like Yersinia pestis, and plague pandemics. *Emerging Infectious Diseases*, *10*(9), pp.1585–1592.

Dudley, J.P., Ginsberg, J.R., Plumptre, A.J., Hart, J.A. and Campos, L.C., 2002. Effects of war and civil strife on wildlife and wildlife habitats. *Conservation Biology*, *16*(2), pp.319–329.

Duplantier, J., Catalan, J., Orth, A., Grolleau, B. and Britton-Davidian, J., 2003. Systematics of the black rat in Madagascar: consequences for the transmission and distribution of plague. *Biological Journal of the Linnean Society*, *78*(3), pp.335–341.

Dutt, A.K., Akhtar, R. and McVeigh, M., 2006. Surat plague of 1994 re-examined. *Southeast Asian Journal of Tropical Medicine and Public Health*, *37*(4), p.755.

Echenberg, M.J., 2002. Pestis redux: The initial years of the third bubonic plague pandemic, 1894–1901. *Journal of World History*, *13*(2), pp.429–449.

Echenberg, M., 2007. *Plague ports: the global urban impact of bubonic plague: 1894–1901*. New York University Press.

Eckert, E.A., 2000. The retreat of plague from Central Europe, 1640–1720: a geomedical approach. *Bulletin of the History of Medicine*, *74*(1), pp.1–28.

Eisen, R.J., Griffith, K.S., Borchert, J.N., MacMillan, K., Apangu, T., Owor, N., Acayo, S., Acidri, R., Zielinski-Gutierrez, E., Winters, A.M. and Enscore, R.E., 2010. Assessing human risk of exposure to plague bacteria in northwestern Uganda based on remotely sensed predictors. *The American Journal of Tropical Medicine and Hygiene*, *82*(5), pp.904–911.

Gage, K.L. and Kosoy, M.Y., 2005. Natural history of plague: perspectives from more than a century of research. *Annual Review of Entomology*, *50*, pp.505–528.

Galvani, A.P. and Novembre, J., 2005. The evolutionary history of the CCR5-Δ32 HIV-resistance mutation. *Microbes and Infection*, *7*(2), pp.302–309.

Gani, R. and Leach, S., 2004. Epidemiologic determinants for modeling pneumonic plague outbreaks. *Emerging Infectious Diseases*, *10*(4), pp.608–614.

Gayer, M., Legros, D., Formenty, P. and Connolly, M.A., 2007. Conflict and emerging infectious diseases. *Emerging Infectious Diseases*, *13*(11), p.1625.

Gilabert, A., Loiseau, A., Duplantier, J.M., Rahelinirina, S., Rahalison, L., Chanteau, S. and Brouat, C., 2007. Genetic structure of black rat populations in a rural plague focus in Madagascar. *Canadian Journal of Zoology*, *85*(9), pp.965–972.

Giles, J., Peterson, A.T. and Almeida, A., 2011. Ecology and geography of plague transmission areas in northeastern Brazil. *PLoS Neglected Tropical Diseases*, 5(1), p.e925.

Gorshkov, O.V., Savostina, E.P., Popov, I., Plotnikov, O.P., Vinogradova, N.A. and Solodovnikov, N.S., 2000. Genotyping Yersinia pestis strains from various natural foci. *Molekuliarnaia genetika, mikrobiologiia i virusologiia*, 3, pp.12–17.

Guiyoule, A., Rasoamanana, B., Buchrieser, C., Michel, P., Chanteau, S. and Carniel, E., 1997. Recent emergence of new variants of Yersinia pestis in Madagascar. *Journal of Clinical Microbiology*, 35(11), pp.2826–2833.

Gupta, M.L. and Sharma, A., 2007. Pneumonic plague, northern India, 2002. *Emerging Infectious Diseases*, 13(4), p.664.

John, T.J., 1994. Learning from plague in India. *The Lancet*, 344(8928), p.972.

John, T.J., 1995. Final thoughts on India's 1994 plague outbreaks. *The Lancet*, 346(8977), p.765.

Joshi, K., Thakur, J.S., Kumar, R., Singh, A.J., Ray, P., Jain, S. and Varma, S., 2009. Epidemiological features of pneumonic plague outbreak in Himachal Pradesh, India. *Transactions of the Royal Society of Tropical Medicine and Hygiene*, 103(5), pp.455–460.

Kellogg, W.H., 1920. An epidemic of pneumonic plague. *American Journal of Public Health*, 10(7), pp.599–605.

Kool, J.L. and Weinstein, R.A., 2005. Risk of person-to-person transmission of pneumonic plague. *Clinical Infectious Diseases*, 40(8), pp.1166–1172.

Kreppel, K.S., Telfer, S., Rajerison, M., Morse, A. and Baylis, M., 2016. Effect of temperature and relative humidity on the development times and survival of Synopsyllus fonquerniei and Xenopsylla cheopis, the flea vectors of plague in Madagascar. *Parasites & Vectors*, 9(1), p.82.

Laudisoit, A., Leirs, H., Makundi, R.H., Van Dongen, S., Davis, S., Neerinckx, S., Lhoest, J. and Libois, R., 2007. Plague and the human flea, Tanzania. *Emerging Infectious Diseases*, 13(5), pp.687–693.

Levi, M.I., Kanatov, Y., Shtelman, A., Minkov, G., Novikova, E. and Optyakov, A.F., 1961. Immunological investigations in plague. 1. Detection of antibodies in sera of experimentally infected animals by means of passive haemagglutination test. *Journal of Microbiology, Epidemiology and Immunobiology USSR*, 32(10), p.1872.

Lien-Teh, W. and Tuck, G.L., 1913a. First report of the North Manchurian plague prevention service. *Journal of Hygiene*, 13(3), pp.237–290.

Lien-Teh, W. and Tuck, G.L., 1913b. Investigations into the relationship of the Tarbagan (Mongolian Marmot) to plague. *The Lancet*, 182(4695), pp.529–535.

Link, V.B., 1955. *A history of plague in the United States of America*. US Government Printing Office.

Lipson, L.G., 1972. Plague in San Francisco in 1900: the United States Marine Hospital Service Commission to study the existence of plague in San Francisco. *Annals of Internal Medicine*, 77(2), pp.303–310.

McCauley, D.J., Salkeld, D.J., Young, H.S., Makundi, R., Dirzo, R., Eckerlin, R.P., Lambin, E.F., Gaffikin, L., Barry, M. and Helgen, K.M., 2015. Effects of land use on plague (Yersinia pestis) activity in rodents in Tanzania. *The American Journal of Tropical Medicine and Hygiene*, 92(4), pp.776–783.

Madan, T.N., 1995. The plague in India, 1994. *Social Science & Medicine*, 40(9), pp.1167–1168.

Mavalankar, D.V., 1995. Indian "plague" epidemic: unanswered questions and key lessons. *Journal of the Royal Society of Medicine*, 88(10), p.547.

Meyer, K.F., 1961. Pneumonic plague. *Bacteriological Reviews*, 25(3), p.249.

Mize, E.L. and Britten, H.B., 2016. Detections of Yersinia pestis east of the known distribution of active plague in the United States. *Vector-Borne and Zoonotic Diseases*, 16(2), pp.88–95.

Neerinckx, S., Bertherat, E. and Leirs, H., 2010. Human plague occurrences in Africa: an overview from 1877 to 2008. *Transactions of the Royal Society of Tropical Medicine and Hygiene*, 104(2), pp.97–103.

Nishiura, H., Schwehm, M., Kakehashi, M. and Eichner, M., 2006. Transmission potential of primary pneumonic plague: time inhomogeneous evaluation based on historical documents of the transmission network. *Journal of Epidemiology & Community Health*, 60(7), pp.640–645.

Parmenter, R.R., Yadav, E.P., Parmenter, C.A., Ettestad, P. and Gage, K.L., 1999. Incidence of plague associated with increased winter-spring precipitation in New Mexico. *The American Journal of Tropical Medicine and Hygiene*, 61(5), pp.814–821.

Perry, R.D. and Fetherston, J.D., 1997. Yersinia pestis – etiologic agent of plague. *Clinical Microbiology Reviews*, 10(1), pp.35–66.

Rahelinirina, S., Duplantier, J.M., Ratovonjato, J., Ramilijaona, O., Ratsimba, M. and Rahalison, L., 2010. Study on the movement of Rattus rattus and evaluation of the plague dispersion in Madagascar. *Vector-Borne and Zoonotic Diseases*, 10(1), pp.77–84.

Rasmussen, S., Allentoft, M.E., Nielsen, K., Orlando, L., Sikora, M., Sjögren, K.G., Pedersen, A.G., Schubert, M., Van Dam, A., Kapel, C.M.O. and Nielsen, H.B., 2015. Early divergent strains of Yersinia pestis in Eurasia 5,000 years ago. *Cell*, 163(3), pp.571–582.

Ratovonjato, J., Rajerison, M., Rahelinirina, S. and Boyer, S., 2014. Yersinia pestis in Pulex irritans fleas during plague outbreak, Madagascar. *Emerging Infectious Diseases*, 20(8), p.1414.

Ratsitorahina, M., Chanteau, S., Rahalison, L., Ratsifasoamanana, L. and Boisier, P., 2000. Epidemiological and diagnostic aspects of the outbreak of pneumonic plague in Madagascar. *The Lancet*, 355(9198), pp.111–113.

Raza, G., Dutt, B. and Singh, S., 1997. Kaleidoscoping public understanding of science on hygiene, health and plague: a survey in the aftermath of a plague epidemic in India. *Public Understanding of Science*, 6(3), pp.247–267.

Ruiz, A., 2001. Plague in the Americas. *Emerging Infectious Diseases*, 7(3 Suppl), p.539.

Schneider, M.C., Najera, P., Aldighieri, S., Galan, D.I., Bertherat, E., Ruiz, A., Dumit, E., Gabastou, J.M. and Espinal, M.A., 2014. Where does human plague still persist in Latin America? *PLoS Neglected Tropical Diseases*, 8(2), p. e2680.

Seal, S.C., 1960. Epidemiological studies of plague in India: 1. The present position. *Bulletin of the World Health Organization*, 23(2–3), p.283.

Signoli, M., Séguy, I., Biraben, J.N., Dutour, O. and Belle, P., 2002. Paleodemography and historical demography in the context of an epidemic. *Population*, 57(6), pp.829–854.

Spyrou, M.A., Tukhbatova, R.I., Feldman, M., Drath, J., Kacki, S., de Heredia, J.B., Arnold, S., Sitdikov, A.G., Castex, D., Wahl, J. and Gazimzyanov, I.R., 2016. Historical Y. pestis genomes reveal the European Black Death as the source of ancient and modern plague pandemics. *Cell Host & Microbe*, 19(6), pp.874–881.

Swanson, M.W., 1977. The sanitation syndrome: bubonic plague and urban native policy in the Cape Colony, 1900–1909. *The Journal of African History*, 18(3), pp.387–410.

Tollenaere, C., Duplantier, J.M., Rahalison, L., Ranjalahy, M. and Brouat, C., 2011. AFLP genome scan in the black rat (Rattus rattus) from Madagascar: detecting genetic markers undergoing plague-mediated selection. *Molecular Ecology*, 20(5), pp.1026–1038.

Tollenaere, C., Rahalison, L., Ranjalahy, M., Rahelinirina, S., Duplantier, J.M. and Brouat, C., 2008. CCR5 polymorphism and plague resistance in natural populations of the black rat in Madagascar. *Infection, Genetics and Evolution*, 8(6), pp.891–897.

Tollenaere, C., Rahalison, L., Ranjalahy, M., Duplantier, J.M., Rahelinirina, S., Telfer, S. and Brouat, C., 2010. Susceptibility to Yersinia pestis experimental infection in wild Rattus rattus, reservoir of plague in Madagascar. *EcoHealth*, 7(2), pp.242–247.

Tumanskii, V., 1957. The classification of variants of the plague microbe. *Journal of Microbiology, Epidemiology and Immunobiology*, 28(5/6), pp.776–780.

Tumanskii, V., 1958. *The bacteriology of plague*. The Bacteriological Foundations of the Diagnosis of Plague.

Vogler, A.J., Chan, F., Nottingham, R., Andersen, G., Drees, K., Beckstrom-Sternberg, S.M., Wagner, D.M., Chanteau, S. and Keim, P., 2013. A decade of plague in Mahajanga, Madagascar: insights into the global maritime spread of pandemic plague. *MBio*, 4(1), pp.e00623–12.

Wood, J.W., Ferrell, R.J. and Dewitte-Avina, S.N., 2003. The temporal dynamics of the fourteenth-century Black Death: new evidence from English ecclesiastical records. *Human Biology*, 75(4), pp.427–448.

World Health Organization, 1997. Human plague in 1995= La peste humaine en 1995. *Weekly Epidemiological Record*, 72(46), pp.344–347.

Xu, L., Stige, L.C., Kausrud, K.L., Ari, T.B., Wang, S., Fang, X., Schmid, B.V., Liu, Q., Stenseth, N.C. and Zhang, Z., 2014. Wet climate and transportation routes accelerate spread of human plague. *Proceedings of the Royal Society of London B: Biological Sciences*, 281(1780), p.20133159.

Yu, H.L. and Christakos, G., 2006. Spatiotemporal modelling and mapping of the bubonic plague epidemic in India. *International Journal of Health Geographics*, 5(1), p.12.

Zabolotny, D.K., 1915. Pulmonary plague in 1910–1911. Otchet russkoy nauchnoy ekspeditsii (Report of the Russian Scientific Expedition), Petrograd.

Zimbler, D.L., Schroeder, J.A., Eddy, J.L. and Lathem, W.W., 2015. Early emergence of Yersinia pestis as a severe respiratory pathogen. *Nature Communications*, 6, p.7487.

8   Weaponized plague and plague surveillance

1346 in Kaffa to the modern day

The first documented and accepted incidence of weaponized plague occurred in 1346, when Genoese traders and sailors were infected with the mBD (pneumonic plague) during the siege of Kaffa (now Feodosia, Ukraine) by the Golden Horde and its leader Janibeg in 1346–1347 (Wheelis 2002). After the Golden Horde's army, which was laying siege to Kaffa, was severely decimated by the mBD, decaying corpses were catapulted over the city walls (see Figure 8.1).

Gabriele de' Mussi's account of this activity in 1347 indicates a clear temporal association between these vaulted corpses and infection of the Genoese traders within the city of Kaffa (Derbes 1966; Wheelis 2002; Barras and Greub 2014). Although little or no evidence suggests corpses can precipitate disease epidemics, funeral home employees are twice as likely to react to tuberculin skin tests as non-embalming personnel (Gershon et al. 1998). This clearly suggests people handling corpses can catch lung or mouth-borne viruses or bacteria as a corpse is moved. Certainly, tubercle bacilli remain viable in cadavers for more than 24 hours, with HIV remaining viable for up to 16 days if a body is maintained at temperatures below 2°C (Conly and Johnston 2005; Demiryürek et al. 2002; Gershon et al. 1998; Jensen et al. 2005).

One can easily imagine then falling bodies, catapulted into Kaffa, exploding on impact and releasing blood, tissue, bacteria and viruses as aerosols into the immediate vicinity of each impact. Within the narrow lanes of Kaffa, these aerosols would linger like dust motes waiting to be inhaled by innocent passers-by.

In October, 1347, a Genoese boat full of sick and dying sailors from Kaffa docked at the port city of Messina, on the Mediterranean island of Sicily, releasing the medieval Black Death upon Europe, killing a third or more of Europeans within four years. Mussi, a contemporary Italian chronicler, suggested people viewed mBD as a sign of the end of the world (Bossak and Welford 2015). De' Mussi went further and suggested mBD ended the Middle Ages and initiated the Renaissance (Herlihy 1997).

Although many people have heard of Kaffa and plague, few know that the Japanese in World War II, according to research conducted by the Pingfan Institute or Water Purification Unit 731, dropped ceramic bomblets containing

Figure 8.1. A city under Mongol siege. From the illuminated manuscript of Rashid
ad-Din's Jami al-Tawarikh (created circa 1307). (Edinburgh University
Library. Public Domain.)

Y. pestis-infected fleas onto several Chinese cities initiating several small outbreaks of bubonic plague (Smiley 2008). In October 1940, the city of Ningbo
was attacked by bombers that dropped bubonic plague-infested fleas and grain
wrapped in strips of cotton onto the city, killing over 500 and creating widespread

panic (Goebel 2006; Keiichi 1995). A similar attack on Jinhua on November 28 failed, as did a subsequent attack on Changde in 1941, in part due to rapid Chinese public health department responses and the fact that bubonic plague attacks require plague-infested fleas to locate hosts (endemic or black rats), kill hosts, and then transfer to humans (Keiichi 1995). This method of biological warfare is too indirect and is easy to defeat as it requires sufficient hosts to be present in both rural and urban areas in close proximity to humans to sustain an epidemic. Although the Third Plague Pandemic is raging today, it is not very lethal, or particularly infectious to humans, nor is it very transmissible.

Although rather unlikely in our contemporary post-Cold War world, the possible use of a biological or chemical terror weapon cannot be ignored. On February 13, 2017, North Korea assassinated Kim Jong Nam, the half-brother of North Korea's president Kim Jong Un in Kuala Lumpur international airport using the chemical nerve agent VX that was wiped across Kim Jong Nam's face by two female assailants (Kaiman and Stiles 2017). The only bioterror attack between 1960–1999 to result in casualties occurred in The Dalles, Oregon area in 1984, when the Rajneeshee cult, seeking to manipulate a local election, contaminated restaurant salad bars with *Salmonella typhimurium*. This resulted in 751 cases of food poisoning (Tucker 1999), although none were fatal.

Frighteningly, between 1960 and 1998, of the 135 known biological or chemical terror incidents, an increasing number were directed against a general civilian population and/or organization, or iconic buildings (Tucker 1999). Certainly, in recent times, chemical weapons have been used in the Iran-Iraq War by Saddam Hussein against Iranians and Iraqi Kurds in Sardasht in 1987 (Lafayette 2002) and in Halabja, Iraq in March 1988, with as many as 5,000 killed and 10,000 injured (BBC News 1988). In 1985, VX and sarin were used by Cuban soldiers working for the Angolan government against South African-supported UNITA forces in Angola (Keim and Burstein 2002). In Syria, chemical weapons including sarin, mustard gas, and chlorine gas have been used indiscriminately by Bashar al-Assad's regime, ISIS, and the Jaysh al-Islam terrorist organization (Warrick 2013). In the deadliest incident, more than 58 died in the Kan Shaykhun sarin attack on April 4, 2017 by Bashar al-Assad's regime (BBC News 2017). The increasingly indiscriminate use of chemical weapons in Angola, Iraq, and Syria does not bode well for the world; while chemical weapons are relatively cheap and easy to manufacture, biological weapons remain expensive and difficult to manufacture. Chemical weapons are also predictable, and typically have limited environmental residency times and limited spatial impact. In contrast, biological weapons are unpredictable, very destructive, and transmissible well beyond a battlefield release point. As a weapon, a biological agent release would kill, injure, and/or petrify a population and have catastrophic economic impacts around the world, as trade and transportation, in all forms, might be curtailed in order to prevent a global pandemic. We would return, however temporarily, to a large-scale world.

Mitigating the impacts and spread of human-to-human transmitted pandemics is difficult, as both SARS, which emerged in Guangdong, China, in November of 2002, and the 2009 Swine Flu illustrated. SARS infected 8,448 people and killed 774 around the world in less than 6 months (Smith 2006; Wang and Jolly 2004). The swine flu that emerged in March and April of 2009 in Mexico and the United States had by May 11 spread to some 30 countries (Smith *et al.* 2009). In California, overall case fatalities were 11% of the 1,088 hospital cases but rose to 18–20% among those 50 years or older (Louie *et al.* 2009). SARS, in particular, gives us an insight into the scale of what biological weapons might achieve economically. SARS cost between \$30-100 billion in lost trade and lost tourism (Hanna and Huang 2004; Smith and Sommers 2003). These indirect costs far exceeded any direct medical costs and were driven by public perceptions of the risk of infection and risk of death from infection, nearly all of which was driven by the fearmongering world press (Smith 2006).

SARS also illustrated just how critical a few hubs in global airline traffic could be and that individual 'superspreaders' could dictate the speed and direction of disease transmission (Bossak and Welford 2010). Hufnagel *et al.* (2004) found that shutting down just 2% of the largest global city airports – those connecting the largest cities –dramatically reduced the spread of a simulated pandemic. However, 27.5% of the principal airline nodes such as Atlanta, Dallas-Fort Worth, Heathrow, and Frankfurt had to be shut down before a similar effect was observed (Hufnagel *et al.* 2004). Later work suggests that 200–300 airports of the world's largest cities would have to be closed to limit a pandemic (Bobashev *et al.* 2008). During SARS, just seven or maybe ten 'superspreaders' moved SARS from Hong Kong neighborhoods to hospitals to hotels and onto international flights to China and Canada (Anderson *et al.* 2004; Cori *et al.* 2009; Shen *et al.* 2004). Air travel even determines the speed, direction, and spatial pattern of annual flu within the US. (Grais *et al.* 2004).

As both SARS and the 2009 swine flu illustrate, pandemics can erupt and spread around the world today at the speed of international airline traffic. This necessitates significant and continuous investment in disease surveillance, anti-bioterrorism, and regional, national, and internationally coordinated efforts to quarantine newly emergent diseases and/or bioterror weapons. For instance, modeling of disease intervention efforts suggests that unless border and internal travel restrictions were 99% effective, a newly emergent influenza pandemic would sweep across the US in 2–3 weeks (Ferguson *et al.* 2006). Other work suggests that if travel restrictions were 95% effective, this would delay the initial spread of a pandemic disease (Epstein *et al.* 2007). However, the spatial heterogeneity of airline traffic connectivity suggests that hub airports would control the speed and spatial extent of a pandemic, while poorly connected cities could avoid infection, or, if such a city was the initial site of infection, reduce the likelihood of the disease becoming pandemic (Colizza *et al.* 2006). More recent simulations suggest the arrival of infectious individuals into countries would be governed by long-range airline traffic (Balcan *et al.* 2009). Either way, travel restrictions would have catastrophic implications for national and global trade. Given the uncontrollability of bioterror weapons and the global economic impacts that a bioterror weapon might unleash, it is highly unlikely that a nation would sanction the use of a bioweapon, but this logic fails with terrorists, particularly those who might embrace death and a belief in the afterlife.

SARS and the 2009 swine flu displayed extremely high transmissibility, but very low lethality. I doubt we would be so lucky if we were attacked using a bio-terror weapon because, on a pandemic optimization fulcrum, bioterror weapons would exhibit very high transmissivity and high lethality, something akin to the medieval Black Death (Bossak and Welford 2010).

The question then is why plague is considered a suitable biological weapon. First, pneumonic plague is an effective biological weapon because it is lethal (with a 90–100% mortality rate; Perry and Fetherston 1997) and transmissible and has the potential to become a pandemic such as in the Second Plague Pandemic/medieval Black Death. However, pneumonic plague exhibits a long latency in the human environment, which is not an ideal choice for a biological weapon. Plague's limited residence time in the natural environment (such as soil) makes it more suitable as a biological weapon than anthrax (Orent 2001, 2004). Nevertheless, the World Health Organization and the CDC in Atlanta catalogue, track, and investigate all plague outbreaks (Orent 2001). Russian bioweapon scientists believe pneumonic plague is so transmissible (and most human populations are susceptible; WHO 1970) that all those exposed to weaponized pneumonic plague would develop plague and then, for good measure, infect others (Orent 2001). Re-analysis of the plague transmissivity among pneumonic plague outbreaks in Mukden, China (1946) and Madagascar (1957) support Russian bioweapon experts' contention that weaponized pneumonic plague would be highly infective (Nishiura *et al.* 2006). Nishiura *et al.* (2006) established that basic reproduction numbers, R_0, ranged from 2.8 to 3.5 for the two epidemics. In other words, each infected individual would infect another 2.8 to 3.5 people. This is comparable to community-based 1918 Spanish Flu basic reproduction numbers of 2.4–4.3 (Vynnycky *et al.* 2007). We should not forget that the 1918 Spanish Flu infected over 500 million and killed in excess of 20 million people, spreading around the globe in less than two years (Patterson and Pyle 1991).

However, pneumonic plague can be treated with streptomycin or gentamicin, or if these are not available, doxycycline, chloramphenicol or ciprofloxacin, and all reduce mortality rates from 40–60% to 10% (Cunha 2002; Greenfield *et al.* 2002; Whitby *et al.* 2002). These antibiotics are cheap and readily available if plague bioweapons are used, and pneumonic plague can also be avoided using surgical masks (Cunha 2002). Nevertheless, a WHO assessment noted that an antibiotic-sensitive pneumonic plague weapon released over cities of 5 million would yield case fatality rates of 10% or 100,000 deaths and 500,000 hospitalizations for developed nations (WHO 1970). Among developing nations with an inadequate supply of antibiotics, 50–70% would die in the initial attack, and 50% would flee the attacked city, leading to secondary infections and another 250,000 deaths (WHO 1970).

If a strain of antibiotic-resistant plague was weaponized and used, something Russian bioweapons scientists were working on before Boris Yeltsin attempted to shut down the Biopreparat program in 1992 (Orent 2001), the medieval Black Death would look like child's play; billions would die. In a terrifying analysis of post-Soviet bioweapon capability, Milton Leitenberg and Raymond Zilinskas argue that research into weaponized antibiotic-resistant plague was still ongoing as of 2012 (Leitenberg *et al.* 2012).

Recent incidences of antibiotic-resistant plague from Madagascar (e.g., Chanteau *et al.* 1998), where pneumonic plague occasionally erupts (e.g., 1959, 1997, 2000; Gani and Leach 2004; Kool and Weinstein 2005; Meyer 1961; Ratsitorahina *et al.* 2000) has led to significant research funding. A multidrug-resistant *Y. pestis* was isolated from a patient in the Ambalavao district of Madagascar in 1995 (Galimand *et al.* 1997). Rather frighteningly, the multidrug resistance was conferred by a conjugative plasmid plP1202 that originated among Enterobacteriaceae such as *E. coli* (Galimand *et al.* 1997). Under laboratory conditions this plasmid easily moved between *E. coli* and *Y. pestis* (Galimand *et al.* 1997). A second strain of multidrug-resistant *Y. pestis* was isolated that yielded a plasmid that conferred resistance to streptomycin, and again this plasmid easily transferred among other *Y. pestis* and *E. coli* strains (Guiyoule *et al.* 2001). In Madagascar in 2007, MLVA-based phylogenies (MLVA or multiple-locus variable number tandem repeat analysis is a method used to perform molecular typing) identified ten plague subclades, of which two were new to science, from among 93 human clinical samples of bubonic and pneumonic plague (Riehm *et al.* 2015). This is truly terrifying – such diversity from just one island including two new subclades. It is little wonder that the WHO and CDC maintain active plague surveillance systems across the globe.

Fortunately, Dr. Ashok Chopra at the University of Texas is leading a team trying to develop a multivalent adenoviral plague vaccine for all three plague forms (Charlton 2017; Sha *et al.* 2016; Tiner *et al.* 2015). Since 2011, this team has been awarded over $1.75 million dollars in funding, with over $1.5 million coming from private, non-government sources (UT System Experts 2017). Over the last 100 years, plague vaccine research using F1.LcrV-based vaccines protected mice but failed to protect monkeys (Smiley 2008). F1 is the *Yersinia pestis* capsular protein, and LcrV is one of *Yersinia pestis'* virulence protein factors (Smiley 2008). Given the low priority government and non-government research funding sources have afforded tropical diseases over the last 30–40 years, it is rather surprising how much funding Dr. Chopra's research group has received. However, given that Dr. Chopra's group is working on the production of one vaccine to combat the three variants of plague, the funding makes sense, especially considering (1) Cold War weaponized plague research, (2) the existence of antibiotic-resistant plague, (3) the rapid accumulation of new plague subclades, and (4) the possibility of bioterrorism.

Until a plague vaccine capable of protecting against all forms and strains of plague is made available, "the best defense against plague remains vigilance" (Orent 2001, p. 5). It is unlikely that Russia will completely abandon research into weaponized plague, nor can we expect other 'rogue' nations to ignore such a suitable biological weapon, or a global vaccine program comparable to smallpox to eliminate plague. Plague in all its forms and strains exists in a variety of animal reservoirs across Asia, its homeland, across Africa, and the Americas. It is only in Europe, which was devastated by the Second Plague Pandemic, where today plague is not endemic. Nevertheless, plague surveillance is critical if we are to continue to suppress another plague pandemic, particularly if a locally eruptive plague is antibiotic-resistant.

Risk management, plague surveillance and modeling

Predictive statistical modeling using consistent, continuous remotely sensed data at a variety of spatial and temporal scales offers public health officials and scientists a cost-effective means to evaluate plague risk across broad domains (Eisen *et al.* 2010). Remote sensing and risk assessment of plague is particularly useful in remote, rural areas in poor countries with underfunded medical and public health systems. However, the spatial scale of remotely sensed data is critical to any predictive modeling of plague risk and success of any follow-up on the ground by public health officials. In other words, the resolution of the spatial data is critical, as is the capability of the model to integrate large volumes of data across large domains (Neerinckx *et al.* 2008). Prior to 2010, GIS-based spatial risk models of plague in Africa were regional or continental in scale, yet plague prevention and control activities (spraying for fleas) in Uganda (and elsewhere across Africa and the world) are typically conducted at the scale of a village or even household after human exposure to plague (Eisen *et al.* 2010; Gage *et al.* 1999). So, prior to 2010, risk modeling of plague could not help risk management of plague because the risk of plague and instances of plague require finer resolution data and modeling down to the sub-village scale (Eisen *et al.* 2010). Since 2010, sub-village scale data resolution in Uganda has offered fine-scale spatial resolution of plague risk and, in future, will allow public health officials to more accurately target places at risk of plague while conserving limited funds (Eisen *et al.* 2010). In Uganda, plague risk is also higher above 1300 m and where fields are left fallow in January (Eisen *et al.* 2010; Winters *et al.* 2009). This supports work in Madagascar, where plague risk is highest in upland areas above 800 m after harvests are collected (Rahelinirina *et al.* 2010). In both upland areas of Uganda and Madagascar where plague remains, diversity among plague-carrying fleas also seems to play a role in sustaining plague (Eisen *et al.* 2012; Miarinjara *et al.* 2016). In Madagascar, three flea species from various rodents carry plague – *Sinopsyllus fonquerniei* and *X. cheopis* that were previously known to carry plague, and for the first time *X. brasiliensis* (Miarinjara *et al.* 2016). The human flea *Pulex irritans* has also been implicated in transmitting plague in Madagascar in 2013 (Ratovonjato *et al.* 2014).

Dry weather between December and February and wet periods prior to harvest also predict strongly for plague (MacMillan *et al.* 2011; Moore *et al.* 2012). Homes in the rural West Nile region of Uganda with pigs and stores of water, corn, and processed foods also predict higher for plague (MacMillan *et al.* 2011), suggesting plague-infected, home-invading rats *target* homes with abundant blood-meals (pigs) and stores of food and water.

In Kazakhstan since 1949, extensive monitoring of plague among gerbils by the Kazakhstan Anti-Plague Research Institute indicates that when a critical threshold of flea-plague abundance among gerbil colonies is reached, plague erupts pandemically among gerbils after a 2-year delay (Davis *et al.* 2004). Such surveillance allows targeted spraying of fleas within gerbil colonies at least a year ahead of any eruptive pandemic threat of plague. Interestingly, spraying of Lambda-Cyhalothrin in north-west Uganda to control fleas and hence plague was found to be effective for up to 100 days per application, and Lambda-Cyhalothrin also controlled mosquitos (Borchert *et al.* 2012). In contrast, host-targeted rodent baiting using Imidacloprid to kill fleas required repeated baiting (Borchert *et al.* 2010). In Kazakhstan, inter-species movement of plague from gerbils to humans seems to occur when yet another threshold is breached (Samia *et al.* 2011). This threshold operates on a year-long lag; if the preceding spring temperatures were 1°C above average, the next year's fall will see a 77% increase in human plague outbreaks (Samia *et al.* 2011). The identification of two distinct plague thresholds – one ecological and one climatological – offer public health officials in Kazakhstan and elsewhere within Central Asia ample opportunity to respond and target gerbil fleas.

Furthermore, ignoring non-rodent vectors of plague could and has proved deadly. For instance, an outbreak of pneumonic plague in Qinghai Province, China in July 2009 originated with a deceased dog (Wang *et al.* 2011).

Concluding book summary

Today, the four or more plague strains that caused the medieval Black Death through human-to-human pneumonic transmission (that are believed to have come from marmots in Central Asia) have become extinct among humans, although we do not know whether these strains and their SNPs are still present within rodent populations in Central Asia (Green and Schmid 2016). Instead, the bubonic plague that erupted in China in the 1850s is now resident among various rodent populations on all continents (excepting Europe and Antarctica) and periodically erupts among humans. Luckily, these rodent-to-human epidemics kill fewer and fewer each year; nevertheless, there is an increasing likelihood that a naturally generated or laboratory-generated antibiotic resistant pneumonic plague, or just another marmot pneumonic plague (seeded by increasing human disruption of the natural environment), could erupt across the world. The consequences of a pneumonic plague, especially one that is antibiotic-resistant, for our highly integrated, highly dependent world would be catastrophic, as SARS, swine flu, and the 1918 Spanish Flu illustrate. We should not forget that pneumonic mBD plague radically altered the social, political, and economic character of Europe between 1347 and 1859. Wealth shifted from southern to northern Europe. Feudalism began its long decline. The growth of public health and disease monitoring (AKA health spying) associated with the expansion of various local, regional, and international quarantine practices helped create the highly bureaucratic governments we live with today. Plague, along with the eruption of Krakatoa, also ushered in our modern communication age. Highly connected trading or pilgrimage towns, cities or ports were devastated repeatedly, while many peripheral hamlets and villages off the beaten track were spared. And let us not forget, millions and millions died in the Justinianic, medieval Black Death, and bubonic plagues.

Certainly, any attempt to restrict the global transmission of a possible future pneumonic antibiotic-resistant plague would necessitate the closure of the majority of airports and maritime ports for an extended period. Other reactions are harder to predict. It is not too far-fetched to consider the lessons of the movie *World War Z* when we consider a possible pneumonic plague pandemic; humans might just have to abandon the continents and retreat to offshore islands for who knows how long! But only the wealthy, the politically powerful, and those prepared to defend those islands, in other words, various military personnel, would be able to retreat to these island safe havens! Those too fragile, too young, and too old, would be left to die on the continents.

References

Anderson, R.M., Fraser, C., Ghani, A.C., Donnelly, C.A., Riley, S., Ferguson, N.M., Leung, G.M., Lam, T.H. and Hedley, A.J., 2004. Epidemiology, transmission dynamics and control of SARS: the 2002–2003 epidemic. *Philosophical Transactions of the Royal Society B: Biological Sciences, 359*(1447), pp.1091–1105.

Balcan, D., Colizza, V., Gonçalves, B., Hu, H., Ramasco, J.J. and Vespignani, A., 2009. Multiscale mobility networks and the spatial spreading of infectious diseases. *Proceedings of the National Academy of Sciences, 106*(51), pp.21484–21489.

Barras, V. and Greub, G., 2014. History of biological warfare and bioterrorism. *Clinical Microbiology and Infection, 20*(6), pp.497–502.

BBC News, 2017. Syria conflict: 'chemical attack' in Idlib kills 58. www.bbc.com/news/world-middle-east-39488539, accessed June 13, 2017.

BBC News On this Day, 1988. Thousands die in Halabja gas attack. BBC News (March 16, 1988). http://news.bbc.co.uk/onthisday/hi/dates/stories/march/16/newsid_4304000/4304853.stm, retrieved May 2017.

Bobashev, G., Morris, R.J. and Goedecke, D.M., 2008. Sampling for global epidemic models and the topology of an international airport network. *PLoS One, 3*(9), p.e3154.

Borchert, J.N., Eisen, R.J., Atiku, L.A., Delorey, M.J., Mpanga, J.T., Babi, N., Enscore, R.E. and Gage, K.L., 2012. Efficacy of indoor residual spraying using lambda-cyhalothrin for controlling nontarget vector fleas (Siphonaptera) on commensal rats in a plague edemic region of northwestern Uganda. *Journal of Medical Entomology, 49*(5), pp.1027–1034.

Borchert, J.N., Enscore, R.E., Eisen, R.J., Atiku, L.A., Owor, N., Acayo, S., Babi, N., Montenieri, J.A. and Gage, K.L., 2010. Evaluation of rodent bait containing imidacloprid for the control of fleas on commensal rodents in a plague-endemic region of northwest Uganda. *Journal of Medical Entomology, 47*(5), pp.842–850.

Bossak, B.H. and Welford, M.R., 2010. Spatio-Temporal attributes of pandemic and epidemic diseases. *Geography Compass, 4*(8), pp.1084–1096.

Bossak, B.H. and Welford, M.R., 2015. Spatio-temporal characteristics of the medieval Black Death. *Spatial analysis in health geography, Part II infectious disease* (pp. 71–84). Ashgate's Geographies of Health Series. London: Ashgate.

Chanteau, S., Ratsifasoamanana, L., Rasoamanana, B., Rahalison, L., Randriambelosoa, J., Roux, J. and Rabeson, D., 1998. Plague, a reemerging disease in Madagascar. *Emerging Infectious Diseases, 4*(1), p.101.

Charlton, C., 2017. Scientists race to develop vaccine for the plague amid fears terrorists could use the medieval disease to kill millions. The Sun, January 13, 2017. www.thesun.

co.uk/news/2599633/scientists-race-to-develop-vaccine-for-the-plague-amid-fears-terrorists-could-use-the-medieval-disease-to-kill-millions/, retrieved May 2017.

Colizza, V., Barrat, A., Barthélemy, M. and Vespignani, A., 2006. The role of the airline transportation network in the prediction and predictability of global epidemics. *Proceedings of the National Academy of Sciences of the United States of America, 103*(7), pp.2015–2020.

Conly, J.M. and Johnston, B.L., 2005. Natural disasters, corpses and the risk of infectious diseases. *Canadian Journal of Infectious Diseases and Medical Microbiology, 16*(5), pp.269–270.

Cori, A., Boëlle, P.Y., Thomas, G., Leung, G.M. and Valleron, A.J., 2009. Temporal variability and social heterogeneity in disease transmission: the case of SARS in Hong Kong. *PLoS Computational Biology, 5*(8), p.e1000471.

Cunha, B.A., 2002. Anthrax, tularemia, plague, ebola or smallpox as agents of bioterrorism: recognition in the emergency room. *Clinical Microbiology and Infection, 8*(8), pp.489–503.

Davis, S., Begon, M., De Bruyn, L., Ageyev, V.S., Klassovskiy, N.L., Pole, S.B., Viljugrein, H., Stenseth, N.C. and Leirs, H., 2004. Predictive thresholds for plague in Kazakhstan. *Science, 304*(5671), pp.736–738.

Demiryürek, D., Bayramoğlu, A. and Ustaçelebi, Ş., 2002. Infective agents in fixed human cadavers: a brief review and suggested guidelines. *The Anatomical Record, 269*(4), pp.194–197.

Derbes, V.J., 1966. de Mussis and the great plague of 1348. *JAMA, 196*(1), pp.59–62.

Eisen, R.J., Borchert, J.N., Mpanga, J.T., Atiku, L.A., MacMillan, K., Boegler, K.A., Montenieri, J.A., Monaghan, A. and Gage, K.L., 2012. Flea diversity as an element for persistence of plague bacteria in an East African plague focus. *PLoS One, 7*(4), p.e35598.

Eisen, R.J., Griffith, K.S., Borchert, J.N., MacMillan, K., Apangu, T., Owor, N., Acayo, S., Acidri, R., Zielinski-Gutierrez, E., Winters, A.M. and Enscore, R.E., 2010. Assessing human risk of exposure to plague bacteria in northwestern Uganda based on remotely sensed predictors. *The American Journal of Tropical Medicine and Hygiene, 82*(5), pp.904–911.

Epstein, J.M., Goedecke, D.M., Yu, F., Morris, R.J., Wagener, D.K. and Bobashev, G.V., 2007. Controlling pandemic flu: the value of international air travel restrictions. *PLoS One, 2*(5), p.e401.

Ferguson, N.M., Cummings, D.A., Fraser, C., Cajka, J.C., Cooley, P.C. and Burke, D.S., 2006. Strategies for mitigating an influenza pandemic. *Nature, 442*(7101), pp.448–452.

Gage, K.L., Gratz, N., Poland, J.D. and Tikhomirov, E., 1999. National health services in prevention and control. In *Plague manual: epidemiology, distribution, surveillance and control* (pp. 167–171). Geneva, Switzerland: World Health Organization.

Galimand, M., Guiyoule, A., Gerbaud, G., Rasoamanana, B., Chanteau, S., Carniel, E. and Courvalin, P., 1997. Multidrug resistance in Yersinia pestis mediated by a transferable plasmid. *New England Journal of Medicine, 337*(10), pp.677–681.

Gani, R. and Leach, S., 2004. Epidemiologic determinants for modeling pneumonic plague outbreaks. *Emerging Infectious Disease Journal, 10*(4), pp.608–614.

Gershon, R.R., Vlahov, D., Escamilla-Cejudo, J.A., Badawi, M., McDiarmid, M., Karkashian, C., Grimes, M. and Comstock, G.W., 1998. Tuberculosis risk in funeral home employees. *Journal of Occupational and Environmental Medicine, 40*(5), pp.497–503.

Goebel, G., 2006. Japan's Unit 731 program. In *Biological weapons*, ed. C. F. Naff. Farmington Hills, MI, USA: Greenhaven Press.

Grais, R.F., Ellis, J.H., Kress, A. and Glass, G.E., 2004. Modeling the spread of annual influenza epidemics in the US: the potential role of air travel. *Health Care Management Science*, 7(2), pp.127–134.

Green, M.H. and Schmid, B., 2016. Plague dialogues: Monica Green and Boris Schmid on plague phylogeny (I). Wordpress. https://contagions.wordpress.com/2016/06/27/plague-dialogues-monica-green-and-boris-schmid-on-plague-phylogeny-i/

Greenfield, R.A., Drevets, D.A., Machado, L.J., Voskuhl, G.W., Cornea, P. and Bronze, M.S., 2002. Bacterial pathogens as biological weapons and agents of bioterrorism. *American Journal of the Medical Sciences*, 323(6), pp.299–315.

Guiyoule, A., Gerbaud, G., Buchrieser, C., Galimand, M., Rahalison, L., Chanteau, S., Courvalin, P. and Carniel, E., 2001. Transferable plasmid-mediated resistance to streptomycin in a clinical isolate of Yersinia pestis. *Emerging Infectious Diseases*, 7(1), p.43.

Hanna, D. and Huang, Y., 2004. The impact of SARS on Asian economies. *Asian Economic Papers*, 3(1), pp.102–112.

Herlihy, D. 1997. *The Black Death and the transformation of the West*. Cambridge, MA, USA: Harvard University Press.

Hufnagel, L., Brockmann, D. and Geisel, T., 2004. Forecast and control of epidemics in a globalized world. *Proceedings of the National Academy of Sciences of the United States of America*, 101(42), pp.15124–15129.

Jensen, P.A., Lambert, L.A., Iademarco, M.F., Ridzon, R. and Centers for Disease Control and Prevention, 2005. Guidelines for preventing the transmission of Mycobacterium tuberculosis in health-care settings, 2005. *MMWR Recommendations and Reports*, 54(RR-17), pp.1–141.

Kaiman, J. and Stiles, M., 2017. Nerve agent was used to kill North Korean leader's half brother, police, say. Los Angeles Times, February 23, 2017. www.latimes.com/world/la-fg-north-korea-kim-assassination-20170223-story.html, retrieved June 13, 2017.

Keiichi, T., 1995. Unit 731 and the Japanese Imperial Army's biological warfare program. In *Japan's wartime medical atrocities: comparative inquiries in science*, eds. J.B. Nie, N. Guo and M. Selden. London: Routledge.

Keim, M. and Burstein, J.L., 2002. Intentional chemical disasters. In *Disaster medicine* (pp. 402–412), eds. D.E. Hogan and J.L. Burstein. New York: Lippincott Williams and Wilkins.

Kool, J.L. and Weinstein, R.A., 2005. Risk of person-to-person transmission of pneumonic plague. *Clinical Infectious Diseases*, 40(8), pp.1166–1172.

Lafayette, L., 2002. Who armed Saddam? World History Archives. www.hartford-hwp.com/archives/51/040.html, accessed September 25, 2017.

Leitenberg, M., Zilinskas, R.A. and Kuhn, J.H., 2012. *The Soviet biological weapons program: a history*. Cambridge, MA, USA: Harvard University Press.

Louie, J.K., Acosta, M., Winter, K., Jean, C., Gavali, S., Schechter, R., Vugia, D., Harriman, K., Matyas, B., Glaser, C.A. and Samuel, M.C., 2009. Factors associated with death or hospitalization due to pandemic 2009 influenza A (H1N1) infection in California. *JAMA*, 302(17), pp.1896–1902.

MacMillan, K., Enscore, R.E., Ogen-Odoi, A., Borchert, J.N., Babi, N., Amatre, G., Atiku, L.A., Mead, P.S., Gage, K.L. and Eisen, R.J., 2011. Landscape and residential variables associated with plague-endemic villages in the West Nile region of Uganda. *The American Journal of Tropical Medicine and Hygiene*, 84(3), pp.435–442.

Meyer, K.F., 1961. Pneumonic plague. *Bacteriological Reviews*, 25(3), p.249.

Miarinjara, A., Rogier, C., Harimalala, M., Ramihangihajason, T.R. and Boyer, S., 2016. Xenopsylla brasiliensis fleas in plague focus areas, Madagascar. *Emerging Infectious Diseases*, 22(12), p.2207.

Moore, S.M., Monaghan, A., Griffith, K.S., Apangu, T., Mead, P.S. and Eisen, R.J., 2012. Improvement of disease prediction and modeling through the use of meteorological ensembles: human plague in Uganda. *PLoS One*, 7(9), p.e44431.

Neerinckx, S.B., Peterson, A.T., Gulinck, H., Deckers, J. and Leirs, H., 2008. Geographic distribution and ecological niche of plague in sub-Saharan Africa. *International Journal of Health Geographics*, 7(1), p.54.

Nishiura, H., Schwehm, M., Kakehashi, M. and Eichner, M., 2006. Transmission potential of primary pneumonic plague: time inhomogeneous evaluation based on historical documents of the transmission network. *Journal of Epidemiology & Community Health*, 60(7), pp.640–645.

Orent, W., 2001. Will the Black Death return? Antibiotic-resistant plague is alive and well. Discover Magazine November, 2001. http://discovermagazine.com/2001/nov/featblack, retrieved May 2017.

Orent, W., 2004. *Plague: the mysterious past and terrifying future of the world's most dangerous disease*. New York: Simon and Schuster.

Patterson, K.D. and Pyle, G.F., 1991. The geography and mortality of the 1918 influenza pandemic. *Bulletin of the History of Medicine*, 65(1), p.4–21.

Perry, R.D. and Fetherston, J.D., 1997. Yersinia pestis – etiologic agent of plague. *Clinical Microbiology Reviews*, 10(1), pp.35–66.

Rahelinirina, S., Duplantier, J.M., Ratovonjato, J., Ramilijaona, O., Ratsimba, M. and Rahalison, L., 2010. Study on the movement of Rattus rattus and evaluation of the plague dispersion in Madagascar. *Vector-Borne and Zoonotic Diseases*, 10(1), pp.77–84.

Ratovonjato, J., Rajerison, M., Rahelinirina, S. and Boyer, S., 2014. Yersinia pestis in Pulex irritans fleas during plague outbreak, Madagascar. *Emerging Infectious Diseases*, 20(8), p.1414.

Ratsitorahina, M., Chanteau, S., Rahalison, L., Ratsifasoamanana, L. and Boisier, P., 2000. Epidemiological and diagnostic aspects of the outbreak of pneumonic plague in Madagascar. *The Lancet*, 355(9198), pp.111–113.

Riehm, J.M., Projahn, M., Vogler, A.J., Rajerison, M., Andersen, G., Hall, C.M., Zimmermann, T., Soanandrasana, R., Andrianaivoarimanana, V., Straubinger, R.K. and Nottingham, R., 2015. Diverse genotypes of Yersinia pestis caused plague in Madagascar in 2007. *PLOS Neglected Tropical Diseases*, 9(6), p.e0003844.

Samia, N.I., Kausrud, K.L., Heesterbeek, H., Ageyev, V., Begon, M., Chan, K.S. and Stenseth, N.C., 2011. Dynamics of the plague–wildlife–human system in Central Asia are controlled by two epidemiological thresholds. *Proceedings of the National Academy of Sciences*, 108(35), pp.14527–14532.

Sha, J., Kirtley, M.L., Klages, C., Erova, T.E., Telepnev, M., Ponnusamy, D., Fitts, E.C., Baze, W.B., Sivasubramani, S.K., Lawrence, W.S. and Patrikeev, I., 2016. A replication-defective human type 5 adenovirus-based trivalent vaccine confers complete protection against plague in mice and nonhuman primates. *Clinical and Vaccine Immunology*, 23(7), pp.586–600.

Shen, Z., Ning, F., Zhou, W., He, X., Lin, C., Chin, D.P., Zhu, Z. and Schuchat, A., 2004. Superspreading SARS events, Beijing, 2003. *Emerging Infectious Diseases*, 10(2), p.256.

Smiley, S.T., 2008. Immune defense against pneumonic plague. *Immunological Reviews*, 225(1), pp.256–271.

Smith, G.J., Vijaykrishna, D., Bahl, J., Lycett, S.J., Worobey, M., Pybus, O.G., Ma, S.K., Cheung, C.L., Raghwani, J., Bhatt, S. and Peiris, J.M., 2009. Origins and evolutionary genomics of the 2009 swine-origin H1N1 influenza A epidemic. *Nature, 459*(7250), pp.1122–1125.

Smith, R.D., 2006. Responding to global infectious disease outbreaks: lessons from SARS on the role of risk perception, communication and management. *Social Science & Medicine, 63*(12), pp.3113–3123.

Smith, R.D. and Sommers, T., 2003. *Assessing the economic impact of public health emergencies of international concern: the case of SARS.* Geneva: World Health Organization.

Tiner, B.L., Sha, J., Kirtley, M.L., Erova, T.E., Popov, V.L., Baze, W.B., van Lier, C.J., Ponnusamy, D., Andersson, J.A., Motin, V.L., Chauhan, S. and Chopra, A.K., 2015. Combinational deletion of three membrane protein-encoding genes highly attenuates Yersinia pestis while retaining immunogenicity in a mouse model of pneumonic plague. *Infection and Immunity, 83*(4), pp.1318–1338.

Tucker, J.B., 1999. Historical trends related to bioterrorism: an empirical analysis. *Emerging Infectious Diseases, 5*(4), p.498.

UT Systems Experts. Chopra, A., 2017. Profile. https://utmb.influuent.utsystem.edu/en/persons/ashok-chopra, downloaded May 30, 2017.

Vynnycky, E., Trindall, A. and Mangtani, P., 2007. Estimates of the reproduction numbers of Spanish influenza using morbidity data. *International Journal of Epidemiology, 36*(4), pp.881–889.

Wang, H., Cui, Y., Wang, Z., Wang, X., Guo, Z., Yan, Y., Li, C., Cui, B., Xiao, X., Yang, Y. and Qi, Z., 2011. A dog-associated primary pneumonic plague in Qinghai Province, China. *Clinical Infectious Diseases, 52*(2), pp.185–190.

Wang, M.D. and Jolly, A.M., 2004. Changing virulence of the SARS virus: the epidemiological evidence. *Bulletin of the World Health Organization, 82*(7), pp.547–548.

Warrick, J., 2013. More than 1,400 killed in Syrian chemical weapons attack, US says. Washington Post, August 30. www.washingtonpost.com/world/national-security/nearly-1500-killed-in-syrian-chemical-weapons-attack-us-says/2013/08/30/b2864662-1196-11e3-85b6-d27422650fd5_story.html?utm_term=.80840223454c, accessed June 13, 2017.

Wheelis, M., 2002. Biological warfare at the 1346 siege of Caffa. *Emerging Infectious Diseases, 8*(9), pp.971–975.

Whitby, M., Ruff, T.A., Street, A.C. and Fenner, F.J., 2002. Biological agents as weapons 2: anthrax and plague. *Medical Journal of Australia, 176*(12), pp.605–608.

Winters, A.M., Staples, J.E., Ogen-Odoi, A., Mead, P.S., Griffith, K., Owor, N., Babi, N., Enscore, R.E., Eisen, L., Gage, K.L. and Eisen, R.J., 2009. Spatial risk models for human plague in the West Nile region of Uganda. *The American Journal of Tropical Medicine and Hygiene, 80*(6), pp.1014–1022.

World Health Organization, 1970. Health aspects of chemical and biological weapons: report of a WHO group of consultants. WHO.

Index

Milton Keynes UK
Ingram Content Group UK Ltd.
UKHW040052071024
449327UK00019B/503